DISCARDED BY
SUFFOLK UNIVERSITY
SAWYER LIBRARY

Finite Difference Equations

H. Levy and F. Lessman

Dover Publications, Inc., New York

This Dover edition, first published in 1992, is an unabridged and unaltered republication of the work first published by The Macmillan Company, New York, in 1961.

Library of Congress Cataloging-in-Publication Data

Levy, H. (Hyman), 1889–
Finite difference equations / H. Levy and F. Lessman.
p. cm.
Originally published: New York : Macmillan, 1961.
Includes index.
ISBN 0-486-67260-3
1. Difference equations. I. Lessman, F.
QA431.L57 1992
515′.625—dc20 92-20438
CIP

Manufactured in the United States by Courier Corporation
67260304
www.doverpublications.com

Preface

THE importance of the Difference Calculus, and its role in the study of Numerical Methods is well established. Just as the subject of Differential Equations grew directly out of the differential calculus to become one of the most important instruments in the hands of mathematical physicists when concerned with continuous processes in nature, so the subject of Difference Equations for the treatment of discontinuous processes is gradually forcing its way to the fore. They arise in the theory of probability, in queueing problems, in the study of electrical networks, in statistical problems, and indeed in all situations in which some sequential relation exists at various discrete values of the independent variable or variables.

The pioneer in this field was undoubtedly Boole who was concerned mainly with linear difference equations and their associated operators. Since then, books on Finite Differences, dealing mainly with Interpolation and allied topics, sometimes devote one or perhaps two chapters to the Linear Equation, and to equations that can be transformed into these. No wider treatment seems so far to have been attempted.* While it is true, as the reader of this book will appreciate, that the whole subject bristles with fundamental problems in Pure Mathematics, it is also true that so far it has not received adequate attention from the pure mathematician. For that reason this book is mainly concerned with methods of solution rather than with a study of these fundamental mathematical issues.

The first four chapters of this book cover largely familiar ground, although there is much that is new in them. Chapter 5, which deals with non-linear equations, and part of Chapter 6 break entirely new ground and have never been published before. In the main we have restricted ourselves to the real variable, because many of the practical problems demand such direct treatment. At the same time the method of treatment lends itself easily to generalization to the complex plane.

We trust that the publication of this book, which embodies much that has been part of M.Sc. courses at the Imperial College for many

* By far the most complete and detailed treatment of the linear equation is that given by L. M. Milne-Thomson in his *Calculus and Finite Differences* (Macmillan).

years, will stimulate others to turn their attention to this increasingly important field. In particular a wide range of investigation is opened up by the discussion in Chapter 5, in which are developed the various modes of expansion of the solutions of non-linear equations, and the conditions under which these are valid. In addition, in Chapter 6, there is the estimation of the eigen-values that arise in connexion with linear difference equations, and the investigation of the properties of the corresponding orthogonal functions in the form of factorial series.

H. LEVY
Emeritus-Professor of Mathematics, University of London;
formerly Head of the Department of Mathematics,
Imperial College of Science and Technology

F. LESSMAN
Lecturer in Mathematics,
Imperial College of Science and Technology

Contents

CHAPTER 1

Elementary Difference Operations

LET $y_1, y_2, y_3, \ldots, y_n, \ldots$ be a sequence of numbers whose order is specified by the subscript n. The number n is therefore an integer and y_n may be regarded as a function of a variable which adopts integral values only. To define such a sequence, that is, to be able actually to write down any specified term or all the terms up to that specified term, it is necessary to know the law or rule of formation. Sometimes the rule *appears* self-evident from the first few terms, as for example with the sequences

1, 3, 5, 7, 9, . . . an arithmetical sequence,
2, 4, 8, 16, . . . a geometrical sequence.

Whether it is justifiable to extract such a rule from a limited set and apply it generally does not concern us at the moment. What interests us is that the sequence is defined for us once we know its starting values and the law of succession. In the first instance the rule is that each member of the sequence is 2 more than the preceding one and that the first member is 1. Thus, in this case, $y_{n+1} - y_n = 2$, $y_1 = 1$. In the second case each member of the sequence is 2 times the preceding member and the first member is 2. Thus here, $y_{n+1} = 2y_n$, $y_1 = 2$. With these combinations between general relation and particular value the sequences are constructible in practice. Moreover it is obvious that in the first case $y_n = 2n - 1$, and in the second case $y_n = 2^n$. Now it should be noticed that these last two expressions for y_n have been obtained more or less at a glance from the sequences as given and on the basis of our past experience of arithmetic and geometric sequences as they occur in elementary algebra. This would not have been seen so easily if these sequences had been stated in the respective forms—

$$y_{n+1} - y_n = 2, \qquad y_1 = 1,$$

and

$$y_{n+1} = 2y_n, \qquad y_1 = 2,$$

both of which equally define their sequences in the sense that by their means they are both constructible.

In this book we are concerned in the first instance with sequences of the form

$$y_1, y_2, y_3, \ldots, y_n, \ldots.$$

We may turn to their study therefore in two ways. We may suppose the sequence is a given set of numbers arranged in an order and we may examine the set to see what kind of regularities show themselves in these numbers. The limited set of numbers is the data, and these are examined and analysed in much the same way as a scientist examines data to discern trends and laws of change. Out of this emerge questions of extrapolation, that is, carrying the sequence beyond the range over which it is given, and questions of interpolation, that is, inserting a value at some point within the sequence at which one of the members is missing just as one might predict the atomic weight of an element from the periodic table when precise experimental knowledge about the element has not yet been derived. This is broadly the one method of approach. The second method, and for us the more important concern of this book, is to turn to a study of the sequence from a general relation of the type

$$y_{n+r} = F(n, y_{n+r-1}, y_{n+r-2}, \ldots, y_n)$$

with its appropriate starting values, and to examine this connexion between successive members for the purpose of exposing other aspects of the structure of the sequence that are of importance.

Let us look a little more closely at this second form of approach if only to see in what sense, if any, the general relation restricts the kind of starting values. Consider

$$y_{n+2} = y_{n+1}^2 - ny_n.$$

Since in this each value of y is determined from a knowledge of the two preceding values, there must be two initial values, say y_1 and y_2, offered as a starting point after which y_3, y_4, etc. can be written down in succession. For example, if $y_1 = 0$ and $y_2 = 1$ then by inserting $n = 1$ in the relation we find $y_3 = 1$, and then with $n = 2$, $n = 3$, $n = 4$, etc. we obtain $y_4 = -1$, $y_5 = -2$, $y_6 = 8$, etc.

Definition of a Difference Equation

1. A relation of the form

$$y_{n+r} = F(n, y_n, y_{n+1}, \ldots, y_{n+r-1})$$

is an *ordinary difference equation of the rth order.*

2. To enable the sequence to be calculable and to be uniquely defined through this relation there is required, in addition, sufficient information to enable r successive values of y, say $y_1, y_2, y_3, \ldots, y_r$, to be initially specified. These are the *boundary conditions* which when used in conjunction with the difference equation make all the succeeding values of y uniquely calculable. That this is adequate for the purpose is obvious if we insert $n = 1$ in the difference equation. This enables us to calculate y_{r+1} from the given boundary values. Then by inserting $n = 2$ we can calculate y_{r+2} from the equation, the given values of y and the calculated value of y_{r+1}. By this process we proceed step by step as far as any later value of y.

We propose to illustrate these statements by means of a well-known elementary problem, viz. given a plane containing n points, no three of which are collinear, to find the number of straight lines that can be formed by joining together every pair of points. The answer, found otherwise, by simple considerations is clearly $\frac{1}{2}n(n-1)$.

Now, let y_n be the number of such lines. Let another point be supposed added to the set, not collinear with any other pair. The number of lines can now be written as y_{n+1}. This can be found from y_n in a very simple way. By adding the $(n + 1)$th point to the set we have in effect made it possible to draw n new lines by joining this new point to each of the previous n points. It follows that we can express this fact by the statement

$$y_{n+1} = y_n + n.$$

This is the difference equation that represents the law of succession for the numbers $y_1, y_2 \ldots$ and enables the calculation to proceed step by step to any given value of n provided the first value y_1 corresponding to $n = 1$ is given. Clearly when $n = 1$ there is no pair of points and therefore no line joining them. Hence the boundary condition is $y_1 = 0$. Accordingly using this value of y_1 for $n = 1$, we calculate in succession from the difference equation, $y_2 = 1$, $y_3 = 3$, $y_4 = 6$, $y_5 = 10$, and so on. To derive the general expression for y_n, viz. $n(n-1)/2$ would involve solving the difference equation in association with the boundary condition $y_1 = 0$. We shall shortly deal with the treatment of such equations for the purpose of deriving the formal expression for y_n; here we need merely remark that the well-known answer to the problem, derived otherwise from elementary considerations, viz. $y_n = n(n-1)/2$, can easily be verified to satisfy the difference equation above and its boundary condition.

The conception of y_n as a function of n where n is an integer is much wider than appears at first sight. From one term to the next the variable increases by unity. Let us suppose, however, that we have a function of the form

$$y_n = f(a + nh)$$

where a and h are constants. Then the sequence $y_0, y_1, y_2, y_3, \ldots, y_n, \ldots$ corresponds to the sequence $f(a)$, $f(a+h)$, $f(a+2h)$, $f(a+3h), \ldots, f(a+nh)$, so that as n increases by unity the function for which y stands has its variable $a + nh$, increasing at each step by the constant h. We call h the *step* or *interval*. In this way we remove the apparent restriction that we are discussing only integral values of the variable. It is, of course, merely a question of a change in scale.

In an earlier paragraph we have pointed out the two lines of approach to the study of a sequence, where in the one case the sequence is given merely in terms of a sample set of numbers $y_1, y_2 \ldots$ and in the other case an adequate number of initial values along with the rule or law of succession expressed through the difference equation. We turn first to a consideration of the case where the sequence is directly given through its sample set of limited extent.

Definition of Δ

The expression $y_{n+1} - y_n$ is called the *difference* of y_n, and symbolically this is written

$$\Delta y_n = y_{n+1} - y_n.$$

It is obvious that the symbol Δ may be seen as an *operator* acting on y_n and is roughly analogous to the differential operator D. We are entitled to call it an *operator* because it does in fact represent a certain operation conducted on y_n in relation to the latter's position in the given sequence of numbers. When this operation is conducted on each y_n, viz. on $y_1, y_2, y_3, \ldots$, that is, when each of these is subtracted from the number that stands on its right, there is created a new sequence of numbers $\Delta y_1, \Delta y_2, \Delta y_3. \ldots$ Let us carry this through in the following simple case—

$$1, 4, 9, 16, 25, 36, 49, \ldots$$

which now stand for

$$y_1, y_2, y_3, \ldots.$$

With our definition of the meaning of the operation,

$$\Delta y_1 = y_2 - y_1 = 4 - 1 = 3, \quad \Delta y_2 = y_3 - y_2 = 9 - 4 = 5,$$
$$\Delta y_3 = y_4 - y_3 = 16 - 9 = 7, \quad \Delta y_4 = y_5 - y_4 = 25 - 16 = 9$$

so that the new sequence

$$\Delta y_1, \Delta y_2, \Delta y_3, \Delta y_4, \ldots$$

now becomes

$$3, 5, 7, 9, \ldots.$$

The original sequence suggests that the general term is n^2. If this is accepted—and it is a gratuitous assumption since we have no knowledge of what happens to the numbers in the original sequence beyond those actually set down—we can say that

$$\Delta y_n = y_{n+1} - y_n = (n + 1)^2 - n^2 = 2n + 1,$$

and this would then provide the general term in the sequence of differences. It is easy to check that this is satisfied by the first few terms derived numerically from the given terms of the original sequence. The symbol Δy_n is usually referred to as the *first difference* in y_n since clearly the operation we have performed on the original sequence can now be repeated on the derived sequence. With the numbers given, the sequence of *second* differences would clearly be

$$5 - 3 = 2, \qquad 7 - 5 = 2, \qquad 9 - 7 = 2,$$

i.e.
$$2, 2, 2, \ldots$$

If as before we assume that $y_n = n^2$ so that $\Delta y_n = 2n + 1$ then

$$\Delta(\Delta y_n) = \Delta(\Delta n^2) = \Delta(2n + 1) = (2n + 3) - (2n + 1) = 2,$$

as is to be expected.

Definition of Δ^r

We propose to use the following symbolism—

$$\Delta(\Delta y_n) = \Delta^2 y_n$$
$$\Delta(\Delta^2 y_n) = \Delta^3 y_n$$
$$\cdots\cdots\cdots\cdots$$
$$\Delta(\Delta^r y_n) = \Delta^{r+1} y_n.$$

From this it is obvious that

$$\Delta^r \Delta^s y_n = \Delta^{r+s} y_n = \Delta^s \Delta^r y_n$$

where r and s are positive integers.

All this may be systematically set out for any given sequence of numbers $y_1, y_2, y_3, \ldots, y_n, \ldots$ in the following table—

$$\begin{array}{cccc}
y_1 & & & \\
 & \Delta y_1 & & \\
y_2 & & \Delta^2 y_1 & \\
 & \Delta y_2 & & \Delta^3 y_1 \\
y_3 & & \Delta^2 y_2 & \\
 & \Delta y_3 & & \Delta^3 y_2 \\
y_4 & & \Delta^2 y_3 & \\
 & \Delta y_4 & & \Delta^3 y_3 \\
\cdot & & \Delta^2 y_4 & \\
 & \cdot & & \Delta^3 y_4 \\
\cdot & & \cdot & \\
 & \cdot & & \cdot \\
y_n & & \cdot & \\
 & \Delta y_n & & \cdot \\
\cdot & & \Delta^2 y_n & \\
 & \cdot & & \Delta^3 y_n
\end{array}$$

EXERCISES (I)

1. Set out the difference table as far as constant values of the differences for the following sequences—

(i) 0, 2, 6, 12, 20, 30, 42, 56, . . .
(ii) 0, 4, 18, 48, 100, 180, 294, . . .
(iii) 1, 3, 19, 85, 261, 631, . . .

2. Assuming that the nth term in each of the above is capable of being represented by a function of n of the second, third, and fourth degrees respectively, find the expression for the nth term in each case and the function of n which $\Delta^2 y_n$ represents.

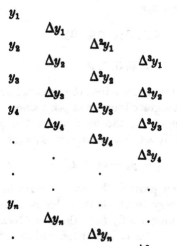

3. (i) If $y_n = \sin 3n$, what is Δy_n?

(ii) Find $\Delta^2 y_n$ if (a) $y_n = a^n \cdot n$

(b) $y_n = n^3 + 1$

(c) $y_n = n^2 - 2n + 3$.

4. (i) Prove that $y_n = 2^n + 3^n$ satisfies the following difference equation—

$$\Delta^2 y_n - 3\Delta y_n + 2y_n = 0.$$

(ii) Prove that $y_n = an(n-1)/2 + n + 1$ satisfies $\Delta^2 y_n = a$ where a is a constant.

(iii) Prove that $y_n = 4^n(An + B)$, where A and B are arbitrary constants, satisfies the equation

$$\Delta^2 y_n - 6\Delta y_n + 9y_n = 0.$$

5. Show that—

(i) $\Delta n^2 = 2n + 1$

(ii) $\Delta n^3 = 3n^2 + 3n + 1$

(iii) $\Delta n^4 = 4n^3 + 6n^2 + 4n + 1$

(iv) $\Delta n(n-1) = 2n$

(v) $\Delta n(n-1)(n-2) = 3n(n-1)$.

6. Perform the following operations—

(i) $(\Delta^2 - \Delta)n(n-1)(n-2)$

(ii) $(\Delta^2 - \Delta)(n^2 + 2n + 1)$.

Some Elementary Properties of Δ

1. The Distributive Law for Δ

It is obvious that Δ satisfies the distributive law of ordinary algebra, for if y_n and z_n are the two nth terms of two sequences,

$$\begin{aligned}\Delta(y_n + z_n) &= (y_{n+1} + z_{n+1}) - (y_n + z_n)\\ &= y_{n+1} - y_n + z_{n+1} - z_n\\ &= \Delta y_n + \Delta z_n.\end{aligned}$$

This can clearly be extended to any finite number of terms, and to an unlimited number if the order of the terms in the unlimited set may be altered without affecting its sum.

Note

Δ must stand before the quantity on which it operates. The expression $y_n\Delta$ has no meaning unless it is followed by a function of n on which it operates.

2. The Commutative Law for Δ

If c is a number which is not a function of n then

$$\begin{aligned}\Delta(cy_n) &= cy_{n+1} - cy_n\\ &= c\Delta y_n,\end{aligned}$$

so that Δ and c may interchange order without affecting the result.

EXERCISE (II)

Show that if $\Delta(c_n y_n) = c_n \Delta y_n$, then $c_{n+1} = c_n$, and c_n is a constant.

3. The Index Law for Δ

In general if r and s are *positive* integers we have the usual rule of exponents of algebra, viz.

$$\Delta^r\Delta^s = \Delta^s\Delta^r = \Delta^{r+s}$$

where it is understood that the function operated on follows the operating symbols. If we assume that this rule is to be valid for $r = 0$, it follows that

$$\Delta^0\Delta^s = \Delta^s$$

or that Δ^0 is equivalent to 1, i.e. $\Delta^0 y_n = y_n$.

It has to be particularly noticed that *so far* no meaning is attached to cases where r and s are negative integers or fractions.

From the definition,

$$\Delta y_n = y_{n+1} - y_n,$$

it follows that

$$\begin{aligned}\Delta^2 y_n = \Delta\,.\,\Delta y_n = \Delta(y_{n+1} - y_n) &= \Delta y_{n+1} - \Delta y_n\\ &= (y_{n+2} - y_{n+1}) - (y_{n+1} - y_n)\\ &= y_{n+2} - 2y_{n+1} + y_n.\end{aligned}$$

In the same way it follows that

$$\begin{aligned}\Delta^3 y_n = \Delta\,.\,\Delta^2 y_n &= \Delta(y_{n+2} - 2y_{n+1} + y_n)\\ &= y_{n+3} - 3y_{n+2} + 3y_{n+1} - y_n,\end{aligned}$$

and it seems clear that it should be possible to express $\Delta^r y_n$ in terms of $y_{n+r}, y_{n+r-1}, \ldots, y_n$. We shall examine this later.

4. The Differences of a Polynomial Function

If
$$y_n = a_0 + a_1 n + a_2 n^2 + \ldots + a_r n^r$$

then
$$\Delta y_n = \sum_{r=0}^{r} \Delta a_r n^r = \sum_{r=0}^{r} a_r[(n+1)^r - n^r]$$

$$= \sum_{r=0}^{r} a_r[rn^{r-1} + \ldots] = \text{function of } (r-1)\text{th degree.}$$

Thus the first difference of a function of the rth degree is a function one degree lower. It follows that the second difference is two degrees lower than the original function. The rth difference is therefore a constant.

We conclude therefore that if a sequence y_n is such that its nth term can be represented by a function of the rth degree the table of $(r+1)$th differences will be zero. Conversely, if the $(r+1)$th differences of a function are zero, then the function may be represented by a polynomial of the rth degree.

Example

n	$y_n = n^2 - n$	Δy_n	$\Delta^2 y_n$	$\Delta^3 y_n$
1	0			
		2		
2	2		2	
		4		0
3	6		2	
		6		0
4	12		2	
		8		0
5	20		2	
		10		
6	30			

EXERCISES (III)

1. Given that $\Delta^3 y_n = 0$ and $y_0 = 1$, $y_1 = 2$, $y_2 = 7$, find the unique expression for y_n in terms of n.

2. If $\Delta^4 y_n = 0$, $y_0 = 1$, $y_1 = 2$, and $y_2 = 6$, find the most general expression for y_n in terms of n.

3. If $\Delta^3 y_n = 0$ and $y_1 = 3$, $y_2 = 5$, $y_3 = 11$, construct the difference table for $n = 0$ to $n = 8$.

5. Linear Operators in Δ

We have shown that the operator Δ obeys the distributive and commutative laws of ordinary algebra. Hence, if α_1, α_2, β_1 and β_2 are constants,

$$\begin{aligned}(\alpha_1 + \beta_1\Delta)(\alpha_2 + \beta_2\Delta)y_n &= \alpha_1(\alpha_2 + \beta_2\Delta)y_n + \beta_1\Delta(\alpha_2 + \beta_2\Delta)y_n \\ &= \alpha_1\alpha_2 y_n + \alpha_1\beta_2\Delta y_n + \beta_1\Delta\alpha_2 y_n + \beta_1\Delta\beta_2\Delta y_n \\ &= \alpha_1\alpha_2 y_n + (\alpha_1\beta_2 + \beta_1\alpha_2)\Delta y_n + \beta_1\beta_2\Delta^2 y_n.\end{aligned}$$

Hence, interchanging α_2 and α_1, and β_2 and β_1,

$$(\alpha_2 + \beta_2\Delta)(\alpha_1 + \beta_1\Delta)y_n = \alpha_2\alpha_1 y_n + (\alpha_2\beta_1 + \beta_2\alpha_1)\Delta y_n + \beta_2\beta_1\Delta^2 y_n.$$

We notice that the right-hand sides of these two relations are identical. Thus we see that the order of subsidiary operators in the total operator $(\alpha_1 + \beta_1\Delta)(\alpha_2 + \beta_2\Delta)$ can be changed provided that the coefficients α_1, β_1, α_2, β_2 are not functions of n. If these coefficients are functions of n, the order of the brackets cannot be changed in general.

This can clearly be extended to an operator involving r such linear factors—

$$(\alpha_1 + \beta_1\Delta)(\alpha_2 + \beta_2\Delta)\ldots(\alpha_r + \beta_r\Delta)y_n$$

the order of which may be changed in any way we choose, as any change can be brought about by a successive interchange of pairs. It follows from what we have said that if $f(z)$ is a polynomial function of degree r and if it be factorized in the form

$$f(z) = \prod_1^r (z - a_r)$$

then
$$f(\Delta)y_n = \prod_1^r (\Delta - a_r)y_n.$$

EXERCISE (IV)

Show that

$$(a_n + b_n\Delta)(c_n + d_n\Delta)y_n = (c_n + d_n\Delta)(a_n + b_n\Delta)y_n$$

if, and only if,

$$d_n/b_n = \Delta c_n/\Delta a_n = A,$$

where A is a constant.

Extension to Functions of a Real Continuous Variable

We commenced this chapter by introducing a sequence

$$y_0, y_1, y_2, \ldots, y_n, \ldots$$

where y_n can be regarded as a function of n, an independent variable which is real but discontinuous and which adopts only integral values. The operator Δ, its definition, and the various elementary properties do not, however, demand this restricted view of the independent variable. In posing our problems in terms of a sequence for which the suffix n necessarily adopts real integral values only, we have unnecessarily restricted the scope of our inquiry.

Suppose that $y(x)$ is a function of a real variable x and that it exists not merely at integral values of x but at all values in a given range, both at and between the integers, then we may define the operator Δ without modification thus—

$$\Delta y(x) = y(x + 1) - y(x)$$

and the rules of operation and the various properties of Δ will be as before.

As the subject develops, certain theorems will be valid for all values of x while others will require that x be restricted as before to integral values only. To avoid the necessity of pointing this out on every occasion we propose to use the convention that when y exists only at integral values of the variable, viz. at $x = n$, we will write it as y_n, a member of a sequence, but when it exists everywhere in any range, then we will write it as $y(x)$, a function of the real continuous variable x.

On this basis it is necessary to look for a moment at what we have said with regard to our definition of an ordinary difference equation. We have called

$$y_{n+r} = F(n, y_n, y_{n+1}, \ldots, y_{n+r-1})$$

a difference equation of the rth order. The form appropriate to x will now be

$$y(x + r) = F[x, y(x), y(x + 1), \ldots, y(x + r - 1)]$$

where the problem posed is to determine the function $y(x)$ satisfying this and some set of boundary conditions. What the necessary and sufficient boundary equations should be to ensure that a unique solution exists is a matter of greater complexity than in the case where x is restricted to n. We shall examine this problem later. For the moment all we are concerned with is stating the convention we propose to adopt for theorems which are valid for x rather than only for n.

Definition of the Operator E^p

We define a new operator which we write in the form E^p such that,

$$E^p y(x) = y(x + p)$$

where p is any number. Thus, E^p operating on a function of a variable x increases that variable by the constant amount p. Since when $p = 0$,

$$E^0 y(x) = y(x)$$

it follows that $E^0 = 1$.

Again, if y_n is the nth member of a sequence so that n is an integer and y_n exists only at such integral values, then,

$$E^p y_n = y_{n+p}.$$

Thus, in this case, E^p has a meaning when p is a positive or negative integer only.

Some Elementary Properties of E^p

1. The Distributive Law

If $y(x)$ and $z(x)$ are two functions of x then

$$E^p[y(x) + z(x)] = y(x + p) + z(x + p) = E^p y(x) + E^p z(x).$$

Thus E^p possesses the distributive property just like an algebraic number.

Once more it must be remembered that E^p, like Δ, must stand before the function on which it operates.

2. The Commutative Law

If c is a number which is not a function of x, then

$$E^p[cy(x)] = cy(x + p) = cE^p y(x).$$

Symbolically therefore,

$$E^p c \equiv cE^p$$

so that the order of c and E^p may be changed without affecting the result of the operation.

3. The Index Law

From the definition, if p and q be any constants

$$E^p \,.\, E^q y(x) = E^p y(x+q) = y(x+p+q)$$
$$= E^{p+q} y(x) = E^q \,.\, E^p y(x).$$

Thus, symbolically,

$$E^p \,.\, E^q \equiv E^{p+q}$$

so that E^p satisfies the ordinary index law in the same way as any algebraic number.

It follows from what we have said that if $f(z)$ is a polynomial function of degree r and if it be factorized in the form $\prod_1^r (z - a_r)$ then

$$f(E)y(x) \equiv \prod_1^r (E - a_r)y(x),$$

where

$$Ey(x) = y(x+1)$$
$$E^2 y(x) = y(x+2)$$
$$\cdots\cdots\cdots\cdots$$
$$E^r y(x) = y(x+r).$$

Notice that the operations corresponding to each of the factors may be carried through singly in succession.

Thus $\quad (E^2 - 1)y(x) = E^2 y(x) - y(x) = y(x+2) - y(x),$

but

$$(E^2 - 1)y(x) = (E-1)(E+1)y(x)$$
$$= (E-1)[y(x+1) + y(x)]$$
$$= (E-1)y(x+1) + (E-1)y(x)$$
$$= y(x+2) - y(x+1) + y(x+1) - y(x)$$
$$= y(x+2) - y(x).$$

EXERCISE (V)

Show that if $y(x+1) = 2y(x)$ then

$$y(x+3) + y(x+2) - 10y(x+1) + 8y(x) = 0.$$

The Relation between E and Δ

It is obvious from the definition and from the fact that E and Δ satisfy the rules of ordinary algebra that a relation must exist between them. In fact since

$$\mathrm{E}y(x) = y(x+1)$$

and

$$\Delta y(x) = y(x+1) - y(x),$$

$$\Delta y(x) = \mathrm{E}y(x) - y(x),$$

$$= (\mathrm{E} - 1)y(x),$$

it follows that the operator Δ is equivalent to the mixed operation E − 1. We write therefore

$$\Delta \equiv \mathrm{E} - 1,$$

where the function to be acted on is for the moment omitted. We are now interested in the study of the operations themselves.

By a rearrangement of the terms we have also

$$y(x+1) = y(x) + \Delta y(x),$$

that is

$$\mathrm{E} \equiv 1 + \Delta.$$

The two symbols related in this simple way may be moved about from one side to the other like algebraic quantities, and therefore ordinary algebraic operations may be performed on them. It is exceedingly important to realize this because it implies that, provided the symbols can be interpreted in actual operations on the function $y(x)$, practically all the theorems of algebra may be applied to these two operators.

In fact

$$f(\mathrm{E})y(x) = f(1+\Delta)y(x)$$

$$= f(1)y(x) + \frac{f'(1)}{1!}\Delta y(x) + \frac{f''(1)}{2!}\Delta^2 y(x) + \ldots.$$

And again

$$f(\Delta)y(x) = f(-1+\mathrm{E})y(x)$$

$$= f(-1)y(x) + \frac{f'(-1)}{1!}\mathrm{E}y(x) + \frac{f''(-1)}{2!}\mathrm{E}^2 y(x) + \ldots$$

$$= f(-1)y(x) + \frac{f'(-1)}{1!}y(x+1)$$

$$+ \frac{f''(-1)}{2!}y(x+2) + \ldots.$$

In both cases the series on the right terminates if $f(\Delta)$ is of finite degree in Δ.

Example

$$f(\Delta)a^x = f(-1)a^x + \frac{f'(-1)}{1!}a^{x+1} + \frac{f''(-1)}{2!}a^{x+2} + \dots$$
$$= a^x\left[f(-1) + a\frac{f'(-1)}{1!} + a^2\frac{f''(-1)}{2!}\dots\right]$$
$$= a^x f(-1 + a).$$

EXERCISE (VI)

Show directly that

$$\Delta^r a^x = (a - 1)^r a^x.$$

Hence deduce the above theorem.

Let $(x)_r$ represent r factors commencing with x and running down in steps of unity. Accordingly,

$$(x)_r = x(x - 1)(x - 2)\dots(x - r + 1).$$

Thus with the usual notation, $n!$ would be written equally well $(n)_n$. Taking first differences,

$$\begin{aligned}\Delta(x)_r &= (x + 1)_r - (x)_r \\ &= (x + 1)x(x - 1)(x - 2)\dots(x - r + 2) \\ &\qquad - x(x - 1)(x - 2)\dots(x - r + 1) \\ &= x(x - 1)(x - 2)\dots \\ &\qquad (x - r + 2)(x + 1 - x + r - 1) \\ &= r(x)_{r-1}.\end{aligned}$$

Thus the rule for differencing $(x)_r$ with respect to x is analogous to the rule for differentiating x^r with respect to x.

It follows that

$$\Delta^2(x)_r = r(r - 1)(x)_{r-2},$$
$$\Delta^3(x)_r = r(r - 1)(r - 2)(x)_{r-3},$$
$$\dots\dots\dots\dots\dots$$
$$\Delta^r(x)_r = r!.$$

EXERCISE (VII)

Show that

$$(E + 3)^5(x)(x - 1)(x - 2) = 4^4x(4x^2 + 3x + 8) + 960.$$

Application

Let $f(a + xh)$ be a polynomial function of x of the rth degree. It must be possible to express $f(a + xh)$ in the form

$$f(a + xh) = A_0 + A_1(x)_1 + A_2(x)_2 + \dots + A_r(x)_r \tag{1}$$

for the function on the right-hand side is itself a polynomial of the rth degree containing $(r + 1)$ arbitrary constants. It can therefore represent any such polynomial. We propose to find the constants $A_0, A_1, A_2, \dots, A_r$.

Inserting $x = 0$ on both sides of (1) it follows that $A_0 = f(a)$. Taking the first difference of both sides and inserting $x = 0$, we find $A_1 = \Delta f(a)$; taking second differences and inserting $x = 0$, we obtain $A_2 = \Delta^2 f(a)/2!$ In the same way, $A_3 = \Delta^3 f(a)/3!$ and $A_r = \Delta^r f(a)/r! = r$th difference of $f(a + xh)$ at $x = 0$. Accordingly,

$$f(a + xh) = f(a) + (x)_1 \Delta f(a)/1! + (x)_2 \Delta^2 f(a)/2! + \dots + (x)_r \Delta^r f(a)/r!$$

which is the form analogous to Taylor's theorem for a polynomial.

Example

To write x^4 in the form

$$x^4 = A_0 + A_1(x)_1 + A_2(x)_2 + A_3(x)_3 + A_4(x)_4.$$

In this case, $x^4 = f(a + xh)$ and we may take $a = 0$ and $h = 1$. The theorem then states,

$$f(x) = f(0) + (x)_1 \Delta f(0)/1! + \dots + (x)_r \Delta^r f(0)/r!$$

Now,

$$\Delta x^4 = 4x^3 + 6x^2 + 4x + 1,$$
$$\Delta^2 x^4 = 4(3x^2 + 3x + 1) + 6(2x + 1) + 4 = 12x^2 + 24x + 14,$$
$$\Delta^3 x^4 = 12(2x + 1) + 24 = 24x + 36,$$
$$\Delta^4 x^4 = 24.$$

Thus,

$$A_0 = f(0) = 0,$$
$$A_1 = \Delta f(0)/1! = 1,$$
$$A_2 = \Delta^2 f(0)/2! = 7,$$
$$A_3 = \Delta^3 f(0)/3! = 6,$$
$$A_4 = \Delta^4 f(0)/4! = 1.$$

Hence,

$$x^4 = x + 7x(x - 1) + 6x(x - 1)(x - 2) + x(x - 1)(x - 2)(x - 3).$$

EXERCISES (VIII)

Write the following polynomials in the form—

$$A_0 + A_1(x) + A_2(x)_2 + \ldots + A_r(x)_r$$

1. $3x^3 - 2x^2 + 6x$
2. $1 - 3x + 4x^2$
3. $1 - 12x + 22x^2 - 12x^3 + 2x^4$.

The simple identical relation between the operators E and Δ, $E = 1 + \Delta$, may be used to derive certain of the formulae we have already obtained and others which will be useful later when we turn to problems of interpolation and extrapolation of functions. For example, the last formula in the previous paragraph, the analogue to Taylor's theorem, is nothing more than the statement that if

$$E = 1 + \Delta,$$

then $$E^x = (1 + \Delta)^x,$$

therefore $$E^x y(0) = (1 + \Delta)^x y(0),$$

$$= (1 + x\Delta + x(x - 1)\Delta^2/2! \ldots)y(0),$$

i.e. $$y(x) = y(0) + x\Delta y(0) + x(x - 1)\Delta^2 y(0)/2! + \ldots.$$

This expresses the value of $y(x)$ in terms of the successive differences of $y(x)$ at $x = 0$. If $y(x)$ is a polynomial this series terminates.

Again, since

$$\Delta = E - 1$$

$$\Delta^n = (E - 1)^n$$

where n is any integer, therefore

$$\Delta^n y_0 = (E - 1)^n y_0$$

$$= [E^n - nE^{n-1} + \ldots + (-1)^n]y_0$$

or $$\Delta^n y_0 = y_n - ny_{n-1} + \ldots + (-1)^n y_0,$$

which is an expression giving the nth difference of y_0 in terms of the values of the terms of the sequence $y_0, y_1, \ldots, y_n, \ldots$.

Both these formulae will be examined later when we come to a study of the Difference Table.

Difference of a Product

$$\begin{aligned}\Delta[u(x) \,.\, v(x)] &= u(x+1) \,.\, v(x+1) - u(x) \,.\, v(x)\\ &= u(x+1)[v(x+1) - v(x)] + v(x)[u(x+1) - u(x)]\\ &= u(x+1)\Delta v(x) + v(x)\Delta u(x),\end{aligned}$$

which is the analogue in the difference calculus to the rule in the differential calculus for differentiating a product.

Illustration

Consider first

$$\begin{aligned}\Delta[a^x y(x)] &= a^{x+1}\Delta y(x) + y(x)(a-1)a^x\\ &= a^x[a\Delta y(x) + (a-1)y(x)]\\ &= a^x(a\Delta + a - 1)y(x).\end{aligned}$$

Now regarding a^x and $(a\Delta + a - 1)y(x)$ as two functions of x whose product is to be differenced, we can repeat the process thus—

$$\Delta^2 a^x y(x) = a^x(a\Delta + a - 1)^2 y(x),$$
$$\Delta^r a^x y(x) = a^x(a\Delta + a - 1)^r y(x),$$

where $(a\Delta + a - 1)^r$ is expansible by the ordinary binomial theorem in order to set out its meaning in simple difference operations.

It follows that if $f(\Delta)$ is a polynomial function of finite degree then

$$f(\Delta)a^x y(x) = a^x f(a\Delta + a - 1)y(x).$$

EXERCISES (IX)

Using the formula for the difference of a product, show that

1. $\Delta(x+1) \,.\, a^x = a^x[(a-1)x + 2a - 1]$.
2. $\Delta u(x)v(x)w(x) = u(x+1)v(x+1)\Delta w(x)$
$\qquad + u(x+1)w(x)\Delta v(x) + w(x)v(x)\Delta u(x)$.
3. $\Delta^2 x^2(x+2) = 6x + 10$.

Difference of a Quotient

$$\begin{aligned}\Delta[u(x)/v(x)] &= u(x+1)/v(x+1) - u(x)/v(x)\\ &= [u(x+1)v(x) - v(x+1)u(x)]/v(x)v(x+1)\\ &= \{[u(x+1) - u(x)]v(x) - u(x)[v(x+1) - v(x)]\}\\ &\qquad\qquad /v(x)v(x+1)\\ &= [v(x)\Delta u(x) - u(x)\Delta v(x)]/v(x)v(x+1).\end{aligned}$$

The similarity with, and the difference from, the corresponding formula in the differential calculus is immediately apparent.

EXERCISES (X)

Using the formula for the difference of a quotient, show that—

1. $\Delta[(x+1)/x] = -1/x(x+1)$.
2. $\Delta[x^2/\sin(2\pi x + a)] = (2x+1)/\sin(2\pi x + a)$.
3. $\Delta[u(x)/a^x] = [u(x+1) - au(x)]/a^{x+1}$.

Difference of a Summation

Let $$S_n = y_1 + y_2 + y_3 + \cdots + y_n$$

so that $$S_{n+1} = y_1 + y_2 + y_3 + \cdots + y_{n+1},$$

then $$\Delta S_n = S_{n+1} - S_n = y_{n+1}.$$

Written in the usual notation for a sum this takes the form

$$\Delta \sum_{r=1}^{n} y_r = y_{n+1} \quad \text{or} \quad \Delta \sum_{1}^{n-1} y_n = y_n.$$

If we remember that the summation sign $\sum_{1}^{n-1}$ can itself be regarded as an operator, this asserts that when Δ acts on $\sum_{1}^{n-1}$ it produces the same effect as if Δ were acting on Δ^{-1} if the latter had an operational meaning and if the index law for Δ applied to negative integers. We should expect therefore that when we seek an interpretation for Δ^{-1} it would be associated with this summation sign.

The Interpretation of $\Delta^{-1}y_n$

We define $\Delta^{-1}y_n$ by the relation $\Delta(\Delta^{-1}y_n) = y_n$.

Let $$z_n = \Delta^{-1}y_n,$$

then $$\Delta z_n = y_n,$$

i.e. $$z_{n+1} - z_n = y_n.$$

Thus also $$z_n - z_{n-1} = y_{n-1}$$

$$\cdots\cdots\cdots\cdots$$

$$z_2 - z_1 = y_1.$$

Adding the left-hand sides and the right-hand sides respectively in the above $(n-1)$ equations, it follows that

$$z_n - z_1 = \sum_1^{n-1} y_n \quad \text{or} \quad \Delta^{-1}y_n = z_1 + \sum_1^{n-1} y_n$$

from the definition of z_n, where z_1 is an arbitrary constant, because it is unspecified.

As a check we notice that

$$\Delta\Delta^{-1}y_n = \Delta z_1 + \Delta \sum_1^{n-1} y_n = \Delta \sum_1^{n-1} y_n = y_n.$$

Thus $\Delta^{-1}y_n = \sum_1^{n-1} y_n +$ constant independent of n, $= \sum^{n-1} y_n$ where the lower limit of the summation is not specified as with the analogous case of the indefinite integral.

EXERCISES (XI)

1. Find y_n when
   (i) $\Delta y_n = \alpha$ (where α is a constant).
   (ii) $\Delta y_n = 2(n+1)$.
   (iii) $\Delta y_n = \alpha^n$ (where α is a constant).
   (iv) $\Delta y_n = n(n-1)$.
   (v) $\Delta y_n = \cos n$.
2. Show from this interpretation of Δ^{-1}, that
   (i) $\Delta^{-2}y_n = \sum_1^{n-1} \sum_1^{n-1} y_n + C_1 n + C_2$,

   where C_1 and C_2 are arbitrary constants.

   (ii) $\Delta^{-3}y_n = \sum^{n-1} \sum^{n-1} \sum^{n-1} y_n + C_1 n(n-1)/2 + C_2 n + C_3$,

   (iii) $\Delta^{-r}y_n = \left(\sum^{n-1}\right)^r y_n + R$,

   where R is an arbitrary polynomial function of n of degree $(r-1)$.

The Commutative Law for Δ and Δ^{-1}

From the definition of Δ and the foregoing interpretation of Δ^{-1}, we have that

$$\begin{aligned}\Delta^{-1}\Delta y_n &= \Delta^{-1}(y_{n+1} - y_n) = A + \sum_1^{n-1} (y_{n+1} - y_n) \\ &= A - y_1 + y_n \\ &= B + \Delta\Delta^{-1}y_n.\end{aligned}$$

Hence $\Delta^{-1}\Delta y_n$ differs from $\Delta\Delta^{-1}y_n$ by an arbitrary constant. This is to be expected as in the case of $\Delta^{-1}\Delta$ the summation occurs at the end of the operation and so introduces an arbitrary constant, whereas in the case of $\Delta\Delta^{-1}$ a difference operation occurs finally and this destroys the arbitrary constant introduced by the summation. We have precisely the same situation in the differential and integral calculus where $D^{-1}Dy(x) - DD^{-1}y(x) = \text{constant}$.

Application to Summation by Parts—Abel's Transformation

Let u_n and v_n be two functions of n.

Let Δ_1, E_1 and Δ_2, E_2 be the Δ and E operators acting only on u_n and v_n respectively. Then,

$$\Delta_1(u_nv_n) = v_n\Delta_1u_n, \qquad E_2(u_nv_n) = u_nv_{n+1}$$

and, as before,

$$\Delta(u_nv_n) = u_{n+1}\Delta_2v_n + v_n\Delta_1u_n,$$

or, written symbolically,

$$\Delta(u_nv_n) = (E_1\Delta_2 + \Delta_1)u_nv_n = E_1\Delta_2(1 + E_1^{-1}\Delta_1\Delta_2^{-1})u_nv_n.$$

Hence $\quad \Delta^{-1}(u_nv_n) = [E_1^{-1}\Delta_2^{-1}/(1 + E_1^{-1}\Delta_1\Delta_2^{-1})]u_nv_n.$

Operating on both sides with $(1 + E_1^{-1}\Delta_2^{-1}\Delta_1)$, we obtain

$$\Delta^{-1}(u_nv_n) = (E_1^{-1}\Delta_2^{-1} - \Delta^{-1}\Delta_1E_1^{-1}\Delta_2^{-1})u_nv_n.$$

Accordingly, on dropping the suffixes,

$$\Delta^{-1}(u_nv_n) = u_{n-1}\Delta^{-1}v_n - \Delta^{-1}(\Delta u_{n-1} \,.\, \Delta^{-1}v_n).$$

This is the equivalent in summations to integration by parts, and can be written—

$$\sum_1^{n-1} u_nv_n = u_{n-1}\sum_1^{n-1} v_n - \sum_{r=1}^{r=n-1} (\Delta u_{r-1} \,.\, \sum_1^{r-1} v_r) + \text{constant},$$

where the summation of the product u_nv_n is expressed in terms of another such summation in which one of the previous terms is differenced and the other summed.

Example

To evaluate $\sum_1^n (n2^n)$.

Let $\qquad u_n = n, \qquad v_n = 2^n,$

then
$$\Delta u_{r-1} = 1,$$
$$\sum_1^{r-1} 2^n = 2^r - 2,$$
$$\sum_1^{n-1} n2^n = (n-1)\sum_1^{n-1} 2^n - \sum_{r=1}^{r=n-1} (2^r - 2) + A$$
$$= (n-2)(2^n - 2) + 2(n-1) + A,$$
or
$$\sum_1^n n2^n = (n-1)(2^{n+1} - 2) + 2n + A.$$

When $n = 1$, the left-hand side is 2, so that $A = 0$.
$$\sum_1^n n2^n = (n-1)2^{n+1} + 2.$$

EXERCISES (XII)

1. Evaluate
$$\sum_1^n n(n-1)(n-2).$$

[Take $u_n = (n-1)(n-2)$, $v_n = n$.]

2. Evaluate
$$\sum_1^{n-1} na^n.$$

Assuming that the operator $(1 + \Delta_1\Delta_2^{-1}E_1^{-1})^{-1}$ may be expanded in the form,
$$(1 + \Delta_1\Delta_2^{-1}E_1^{-1})^{-1} = 1 - \Delta_1\Delta_2^{-1}E_1^{-1} + \Delta_1^2\Delta_2^{-1}E_1^{-2} + \ldots$$
then,
$$\Delta^{-1}(u_nv_n) = E_1^{-1}\Delta_2^{-1}(1 + E_1^{-1}\Delta_1\Delta_2^{-1})^{-1}(u_nv_n)$$
$$= (E_1^{-1}\Delta_2^{-1} - E_1^{-2}\Delta_2^{-2}\Delta_1 + E_1^{-3}\Delta_1^2\Delta_2^{-3} \ldots)u_nv_n,$$
i.e.
$$\Delta^{-1}(u_nv_n) = u_{n-1}\Delta^{-1}v_n - \Delta u_{n-2}\Delta^{-2}v_n + \Delta^2 u_{n-3}\Delta^{-3}v_n - \ldots + A,$$

the terms on the right representing multiple summations of v_n multiplied by successively higher differences of u_n. If u_n is a polynomial the expression on the right terminates.

Interpretation of $(1 - \lambda\Delta)^{-1}y_n$

We define $(1 - \lambda\Delta)^{-1}y_n$ by the relation
$$(1 - \lambda\Delta)(1 - \lambda\Delta)^{-1}y_n \equiv y_n \qquad (2)$$

of which the previous definition of $\Delta^{-1}y_n$ can be shown to be a particular case.

Consider the series

$$y_n + \lambda\Delta y_n + \lambda^2\Delta^2 y_n + \ldots \tag{3}$$

and assume it is either absolutely convergent or finite in extent. The latter case will occur if y_n is a polynomial in n. Operate on this series with $(1 - \lambda\Delta)$. Since (3) is absolutely convergent this operation is legitimate as it involves nothing more than altering the arrangement of the terms in the series. Now,

$$(1 - \lambda\Delta)(y_n + \lambda\Delta y_n + \lambda^2\Delta^2 y_n + \ldots + \lambda^r\Delta^r y_n + R_r)$$

$$\left(\text{where } R_r = \sum_{s=r+1}^{\infty} \lambda^s\Delta^s y_n\right)$$

$$\begin{aligned}
&= y_n + \lambda\Delta y_n + \lambda^2\Delta^2 y_n + \ldots + \lambda^r\Delta^r y_n + R_r \\
&\quad - \lambda\Delta y_n - \lambda^2\Delta^2 y_n - \ldots - \lambda^r\Delta^r - \lambda^{r+1}\Delta^{r+1}y_n - \lambda\Delta R_r \\
&= y_n - \lambda^{r+1}\Delta^{r+1}y_n + R_r - \lambda R_{r+1} + \lambda R_r \\
&= y_n + \text{a remainder which tends to zero as } r \text{ increases,} \\
&\qquad \text{since } R_r \to 0 \text{ and } \lambda^r\Delta^r y_n \to 0 \text{ as } r \to \infty.
\end{aligned}$$

Hence, if

$$(1 - \lambda\Delta)z_n = y_n,$$

a value for z_n is obtained by expanding $(1 - \lambda\Delta)^{-1}$ by the binomial expansion and operating on y_n.

This is not the only solution or the most general solution of the equation in z_n

$$(1 - \lambda\Delta)z_n = y_n,$$

as we shall see shortly when we deal with the actual solution of difference equations, but it is a unique expression, given y_n, which we will take as the unique meaning to be attached to $(1 - \lambda\Delta)^{-1}y_n$.

On this basis we can now inquire whether the operators $(1 - \lambda\Delta)$ and $(1 - \lambda\Delta)^{-1}$ are commutative. Now

$$\begin{aligned}
(1 - \lambda\Delta)^{-1}(1 - \lambda\Delta)y_n &= (1 - \lambda\Delta)^{-1}(y_n - \lambda\Delta y_n) \\
&= (y_n - \lambda\Delta y_n) + \lambda\Delta(y_n - \lambda\Delta y_n) \\
&\qquad + \lambda^2\Delta^2(y_n - \lambda\Delta y_n) + \ldots \\
&= y_n - \lambda\Delta y_n + \lambda\Delta y_n - \lambda^2\Delta^2 y_n \\
&\qquad + \lambda^2\Delta^2 y_n - \ldots \\
&= y_n,
\end{aligned}$$

on the assumption we have made that the series (3) is absolutely convergent. Thus

$$(1 - \lambda\Delta)^{-1}(1 - \lambda\Delta)y_n \equiv (1 - \lambda\Delta)(1 - \lambda\Delta)^{-1}y_n.$$

EXERCISES (XIII)

Perform the following operations—

1. $\dfrac{1}{1 - 3\Delta}\, a^n$ where a is a constant.

2. $\dfrac{1}{2 + \Delta}\, n^2.$

3. $\dfrac{\Delta^2}{1 - 4\Delta}\,(n^4 + 5n^2 + 1).$

In the foregoing elementary discussion we have shown in the first place that the operator Δ behaves like an ordinary algebraic symbol and that provided a unique and clearly defined meaning is attached to the results of operations involving Δ—meanings which force themselves upon us from the actual operations in order that they may provide arithmetically intelligible results—we can apply familiar elementary theorems of algebra to them without contradiction. The formal establishment of these theorems for the operator Δ would proceed step-by-step, with the appropriate interpretations, with the formal establishment of the algebraic theorems themselves. Thus we have seen Δ following the laws of distribution and of indices, and of expansion of a binomial operator by the binomial theorem for a positive integral index. We have seen how a binomial operator can also be formally expanded in this way for an index -1. It is thereafter a mere matter of following step-by-step the formal proofs in elementary algebra to show that a binomial expression of the type $1 + \lambda\Delta$ when raised to *any* index may be expanded in the usual way. We do not propose here to carry through these formal proofs in every case, but rather to assume, without further consideration, that unless there is distinct evidence to the contrary in any given case, that the usual algebraic theorem can be applied. We shall go even further and introduce certain well-known functions, such as the logarithmic and the exponential functions, in operational form, but in each case we shall indicate precisely what consistent interpretation is to be applied to them as operating functions.

EXERCISES (XIV)

1. Show that if for any r

$$(1 + \lambda\Delta)^r y_n = y_n + r\lambda\Delta y_n + \frac{r(r-1)}{2!}\lambda^2\Delta^2 y_n + \ldots,$$

then $$(1 + \lambda\Delta)^r . (1 + \lambda\Delta)^s y_n = (1 + \lambda\Delta)^{r+s} y_n.$$

2. Show that $\Delta^r(a^n y_n) = a^n(a - 1 + a\Delta)^r y_n$ where r is a positive or negative integer. Hence deduce that if $f(z)$ is expansible in a power series of z that $f(\Delta)y_n = a^n f(a - 1 + a\Delta) . a^{-n} y_n$.

3. Show that $(\Delta - \lambda)^r y_n = (1 + \lambda)^{n+r}\Delta^r[y_n/(1 + \lambda)^n]$.

4. Show that $(\mathrm{E} - b)^r y_n = b^{n+r}\Delta^r(y_n/b^n)$.

5. Evaluate $(1 + \Delta^2)^{-1}(n^2 + n + 1)$, $\quad(\Delta^2 + \Delta - 2)^{-1}(n^3 + 1)$.

6. Show that $(1 + \lambda\Delta)^{-1} y_n = (a - 1)a^{-n}\sum_1^{n-1} a^n y_n + Aa^{-n}$ where $\lambda = a/(a - 1)$, $a \neq 1$, and A is an arbitrary constant. Deduce that $\sum_1^\infty n^2/2^n = 6$.

7. Determine $y(x)$ to satisfy $(1 + \Delta)^r y(x) = x^2$, where r is any number.

Some Illustrations of the Use of Symbolic Operators

1. Expression for the Sum of Finite Set of Terms

Any algebraic transformation that does not involve fractional powers can be applied to the operators E and Δ. For example, consider

$$1 + z + z^2 + \ldots + z^{n-1} = \frac{z^n - 1}{z - 1}.$$

Now let $z - 1 = t$, then

$$\frac{z^n - 1}{z - 1} = \frac{(t + 1)^n - 1}{t} = n + \frac{n(n-1)}{2!}t + \frac{n(n-1)(n-2)}{3!}t^2 + \ldots + t^{n-1}.$$

There is therefore the algebraic equivalence—

$$1 + z + z^2 + \ldots + z^{n-1}$$
$$= n + \frac{n(n-1)}{2!}t + \frac{n(n-1)(n-2)}{3!}t^2 + \ldots + t^{n-1},$$

which arises from $t = z - 1$.

Let $$z \equiv \mathrm{E}.$$

Then $$t \equiv \Delta,$$

and

$$1 + E + E^2 + \ldots + E^{n-1}$$

$$= n + \frac{n(n-1)}{2!}\Delta + \frac{n(n-1)(n-2)}{3!}\Delta^2 + \ldots + \Delta^{n-1}.$$

Let each side operate on y_1 then

$$y_1 + y_2 + y_3 + \ldots + y_n = \sum_1^n y_n$$

$$= ny_1 + \frac{n(n-1)}{2!}\Delta y_1 + \frac{n(n-1)(n-2)}{3!}\Delta^2 y_1 + \ldots + \Delta^{n-1}y_1.$$

Therefore, if the differences of y_1 above a certain order are zero, the sum of the n terms of the series $\sum_1^n y_n$ will be expressed in a closed form, the *number* of whose terms is independent of n.

Example

To sum $\sum_1^n n^3 = 1^3 + 2^3 + \ldots + n^3$

Here $y_1 = 1$, $y_2 = 8$, $y_3 = 27$, $y_4 = 64$, $\Delta y_1 = 7$, $\Delta y_2 = 19$, $\Delta y_3 = 37$, $\Delta^2 y_1 = 12$, $\Delta^2 y_2 = 18$, $\Delta^3 y_1 = 6$, and $\Delta^4 y_1 = 0$ etc.

Thus using the formula we have established

$$\sum_1^n n^3 = n + \frac{n(n-1)}{2!} \times 7 + \frac{n(n-1)(n-2)}{3!} \times 12 + \frac{n(n-1)(n-2)(n-3)}{4!} \times 6 = \left[\frac{n(n+1)}{2}\right]^2$$

EXERCISES (XV)

1. If $z = 1 + t$, find the form in t equivalent to

$$1 + 2z + 3z^2 + \ldots + (n+1)z$$

and hence find an expression in terms of differences of y_0 for the sum of

$$y_0 + 2y_1 + 3y_2 + \ldots + ny_{n-1} + (n+1)y_n.$$

2. Evaluate $\sum_1^n n^4$.

3. Evaluate $1 . 2 + 2 . 2 . 3 + 3 . 3 . 4 + \ldots + n^2(n+1)$.

4. Show that

$$f(a - nh) = f(a) - n\Delta f(a-h) + \frac{n(n-1)}{2!}\Delta^2 f(a-2h) + \ldots + \frac{n(n-1)\ldots(n-r+1)}{r!}\Delta^r + (a - rh).$$

5. Show that

$$\Delta^n f(a) = f(a + nh) - nf[a + (n - 1h)] + \frac{n(n-1)}{2!} f[a + (n - 2h)] \ldots.$$

2. Montmort's Theorem on Infinite Summation

Write $\mathrm{E}a_n = a_{n+1}$, $\Delta a_n = a_{n+1} - a_n$,

then
$$\begin{aligned} S &\equiv a_0 + a_1 x + a_2 x^2 + \ldots + a_r x^r + \ldots \\ &= (1 + x\mathrm{E} + x^2\mathrm{E}^2 + \ldots)a_0 \\ &= (1 - x[1 + \Delta])^{-1} a_0 \\ &= \frac{1}{1-x}\left[1 - \frac{x}{1-x}\Delta\right]^{-1} a_0 \\ &= \frac{1}{1-x}\left[1 + \frac{x}{1-x}\Delta + \frac{x^2}{(1-x)^2}\Delta^2 + \ldots\right] a_0 \\ &= a_0/(1-x) + x\Delta a_0/(1-x)^2 + x^2\Delta^2 a_0/(1-x)^3 + \ldots \end{aligned}$$

where the original series must be absolutely convergent. Hence, if in S the coefficients a_r can be expressed as polynomials in r of finite degree m say, then $\Delta^M a_0$ is zero for all $M > m$ and a finite expression for the series S will have been found. This theorem was first discovered by Montmort (*Phil. Trans. R.S.L.*, 1717).

If the coefficients a_r are not expressable as polynomials but their differences diminish rapidly, then in certain cases Montmort's formula is convenient for numerical approximation.

EXERCISES (XVI)

1. Express the following series in finite terms for the ranges of x in which they converge—

(i) $1 + 3x + 5x^2 + 7x^3 + \ldots.$

(ii) $1 \, . \, 2 + 2 \, . \, 3x + 3 \, . \, 4x^2 + 4 \, . \, 5x^3 + \ldots.$

2. Show that if $a_0 > a_1 > a_2 \ldots > 0$, then

$$a_0 - a_1 + a_2 - a_3 + \ldots = a_0/2 - \Delta a_0/2^2 + \Delta^2 a_0/2^3 + \ldots.$$

We have referred to the possibility of using the logarithmic, exponential and other functions in symbolic form. Now let us consider

$$f(x) = 1 + x/1! + x^2/2! + \ldots + x^r/r! + \ldots.$$

The series on the right-hand side can easily be shown to be absolutely convergent. It is a matter of simple rearrangement of terms thereafter to establish the relation

$$f(x) \, . \, f(y) = f(x + y),$$

and to deduce that the series in question is a certain constant raised to the power x. This constant is then represented by e. What is done in relation to these algebraic symbols x and y could equally well have been carried through with the symbols E and Δ. All the theorems applicable to e^x other than those that must involve differentiation and integration will formally be applicable to series such as

$$1 + E/1! + E^2/2! + E^3/3! + \dots$$

which we can therefore write as e^E provided the series, derived before and after the operations are performed on the operand, satisfy the necessary convergence conditions. The derivation of the logarithmic series follows also as a mere matter of algebraic reorganization and it becomes again legitimate to write in a formal sense

$$\log (1 + \Delta) = \Delta - \Delta^2/2 + \Delta^3/3 + \dots$$

provided, of course, that the operations referred to can be carried through, that is to say, provided they are interpretable in terms of the members of the sequences and that the series to which they give rise are of the requisite degree of convergence.

Example

$$\log (1 - \Delta) = -\Delta - \frac{\Delta^2}{2} - \frac{\Delta^3}{3} - \dots$$

$$\log (1 - \Delta) = \log (2 - E) = \log 2 + \log (1 - E/2)$$

$$= \log 2 - \frac{E}{2} - \frac{E^2}{2 \cdot 2^2} - \frac{E^3}{3 \cdot 2^3} - \dots$$

Operate with these equivalent expressions on y_0 where $Ey_n = y_{n+1}$, then

$$y_1/2 + y_2/2 \cdot 2^2 + y_3/3 \cdot 2^3 + \dots$$
$$= y_0 \log 2 + \Delta y_0 + \Delta^2 y_0/2 + \Delta^3 y_0/3 + \dots,$$

where y_0 is determined from y_n which specifies the structure of the first series. For example, if $y_n = n^2 - 1$ then $y_0 = -1$, $y_1 = 0$. $y_2 = 3$, $y_3 = 8$, $\Delta y_0 = 1$, $\Delta^2 y_0 = 2$, $\Delta^3 y_0 = \Delta^4 y_0 = 0$. Thus—

$$\frac{1 \cdot 3}{2 \cdot 2^2} + \frac{2 \cdot 4}{3 \cdot 2^3} + \frac{3 \cdot 5}{4 \cdot 2^4} + \dots = -\log 2 + 2.$$

3.

Let $S = a_0 + a_1x/1! + a_2x^2/2! + a_3x^3/3! + \ldots$

be an absolutely convergent series for some range of values of x. If E be defined by the relation

$$\mathrm{E}a_n = a_{n+1}$$

then $S = a_0 + x\mathrm{E}a_0/1! + x^2\mathrm{E}^2a_0/2! + x^3\mathrm{E}^3a_0/3! + \ldots$

$$= (1 + x\mathrm{E}/1! + x^2\mathrm{E}^2/2! + x^3\mathrm{E}^3/3! + \ldots)a_0.$$

This operational expression we may write as $e^{x\mathrm{E}}$ or, if

$$\Delta a_r = a_{r+1} - a_r$$

$$S = e^{x\mathrm{E}}a_0 = e^{x(1+\Delta)}a_0 = e^x . e^{x\Delta}a_0$$

$$= e^x(1 + x\Delta/1! + x^2\Delta^2/2! + x^3\Delta^3/3! + \ldots)a_0$$

$$= e^x(a_0 + x\Delta a_0/1! + x^2\Delta^2a_0/2! + x^3\Delta^3a_0/3! + \ldots).$$

If a_n is a polynomial function of n of the rth degree then the terms on the right will proceed as far as $x^r\Delta^ra_0/r!$ since all higher differences are zero.

Example

$$S = 1 + 3x/1! + 5x^2/2! + 7x^3/3! + \ldots$$

Here $a_0 = 1,\quad \Delta a_0 = 2,\quad \Delta^2a_0 = \Delta^3a_0 = \ldots = 0,$

so that $S = e^x(1 + 2x).$

EXERCISES (XVII)

Show that

1. $1 + \dfrac{2x}{1!} + \dfrac{9x^2}{2!} + \dfrac{28x^3}{3!} + \dfrac{65x^4}{4!} + \ldots = e^x[1 + x + 3x^2 + x^3].$
2. $3x/1! + 8x^2/2! + 15x^3/3! + \ldots = xe^x(x + 3).$
3. $x/1! + 4x^2/2! + 10x^3/3! + 20x^4/4! + 35x^5/5! + \ldots = xe^x(1 + x + x^2/6).$

4. Application to a Compounded Series

Let $f(x) = f_0 + f_1x/1! + f_2x^2/2! + \ldots,$

and $g(x) = g_0 + g_1x/1! + g_2x^2/2! + \ldots.$

Consider the series

$$S = f_0g_0 + f_1g_1x/1! + f_2g_2x^2/2! + \ldots$$

which we assume is absolutely convergent. Let E be an operator such that

$$\mathrm{E}f_n = f_{n+1}$$

so that

$$\Delta f_n = f_{n+1} - f_n,$$

then

$$\begin{aligned} S &= (g_0 + g_1 x\mathrm{E}/1! + g_2 x^2\mathrm{E}^2/2! + \ldots)f_0 \\ &= g(x\mathrm{E})f_0 \\ &= g(x + x\Delta)f_0 \\ &= g(x)f_0 + \frac{x}{1!}g'(x)\Delta f_0 + \frac{x^2}{2!}g''(x)\Delta^2 f_0 + \ldots, \end{aligned}$$

or alternatively

$$= f(x)g_0 + \frac{x}{1!}f'(x)\Delta g_0 + \frac{x^2}{2!}f''(x)\Delta^2 g_0 + \ldots.$$

Examples

1. It is required to sum the series

$$\frac{3x}{1!} - \frac{7x^3}{3!} + \frac{11x^5}{5!} - \ldots = S \text{ (say)}.$$

Let

$$g(x) = \sin x = x - \frac{x^3}{3!} + \frac{x^5}{5!} - \ldots.$$

Thus $g_0 = 0,\ g_1 = 1,\ g_2 = 0,\ g_3 = -1,\ g_4 = 0,\ g_5 = 1.$

Now from S

$f_0 g_0 = 0,\ f_1 g_1 = 3,\ f_2 g_2 = 0,\ f_3 g_3 = -7,\ f_4 g_4 = 0,\ f_5 g_5 = 11.$

Hence $\quad f_1 = 3,\ f_3 = 7,\ f_5 = 11,\ f_7 = 15, \ldots,$

where f_0, f_2, f_4, f_6 are arbitrary.

Accordingly we may take

$$f_0 = 1,\ f_1 = 3,\ f_2 = 5,\ f_3 = 7,\ f_4 = 9,\ f_5 = 11 \text{ etc.},$$

in which case $\Delta f_0 = 2,\ \Delta^2 f_0 = 0 = \Delta^3 f_0 = \Delta^5 f_0$ etc.

Then

$$\begin{aligned} S &= \frac{3x}{1!} - \frac{7x^3}{3!} + \frac{11x^5}{5!} - \ldots \\ &= g(x)f_0 + \frac{x}{1!}g'(x)\Delta f_0 \\ &= \sin x + \frac{x}{1!}\cos x \,.\, 2 \\ &= \sin x + 2x\cos x. \end{aligned}$$

2. If $f(x) \equiv g(x)$ we have the result that if

$$f(x) = f_0 + f_1^2 x/1! + f_2^2 x^2/2! + \ldots$$
$$= f(x)f_0 + xf'(x)\Delta f_0/1! + x^2 f''(x)\Delta^2 f_0/2! + \ldots.$$

3. To sum the series

$$1 + \frac{3^2 x}{1!} + \frac{5^2 x^2}{2!} + \frac{7^2 x^3}{3!} + \ldots = S$$

It has already been shown (*see* p. 29) that

$$(2x + 1)e^x = 1 + \frac{3x}{1!} + \frac{5x^2}{2!} + \frac{7x^3}{3!} + \ldots.$$

So choose $$g(x) = (2x + 1)e^x,$$

and write $$S = f_0 g_0 + f_1 g_1 \frac{x}{1!} + f_2 g_2 \frac{x^2}{2!} + \ldots$$

$$= f_0 g(x) + \frac{xg'(x)}{1!}\Delta f_0 + \ldots$$

Then $$f_0 = 1, f_1 = 3, f_2 = 5, f_3 = 7, \ldots.$$

So $$\Delta f_0 = 2, \Delta^2 f_0 = \Delta^3 f_0 = \ldots = 0,$$

thence $$S = (2x + 1)e^x + 2xe^x(2x + 3)$$
$$= e^x(4x^2 + 8x + 1).$$

EXERCISES (XVIII)

Sum the following series—

1. $1 - 4x^2/2! + 7x^4/4! - 10x^6/6! + \ldots.$

2. $1 + 3x + \frac{7x^2}{2!} + \frac{13x^3}{4!} + \frac{21x^4}{5!} + \frac{31x^5}{5!} + \ldots.$

3. $3 + \frac{5x^2}{2!} + \frac{7x^4}{4!} + \frac{9x^6}{6!} + \ldots.$

5. Application to the Reversion of Series

Suppose $$u(t) = v(t) + \lambda v(t + 1) + \lambda^2 v(t + 2) + \ldots$$
$$= \sum_{n=0}^{\infty} \lambda^n v(t + n),$$

to determine a function $v(t)$ that satisfies this condition when this series is convergent. This will be satisfied if $|\lambda| < m/M$ and $v(t)$ always remains within fixed finite bounds, viz. $m < v(t) < M$.

Hence $$u(t) = (\sum_0^\infty \lambda^n \mathrm{E}^n)v(t) = \frac{1}{1 - \lambda\mathrm{E}} \cdot v(t).$$

Hence $$v(t) = (1 - \lambda\mathrm{E})u(t) = u(t) - \lambda u(t + 1),$$

a solution that is easily verified by direct substitution. The extent of the generality of this solution will become apparent when we turn to difference equations.

Suppose

$$\begin{aligned} u(t) &= v(t) + \frac{\lambda}{1!}v(t + 1) + \frac{\lambda^2}{2!}v(t + 2) + \ldots, \\ &= \left(1 + \frac{\lambda\mathrm{E}}{1!} + \frac{\lambda^2\mathrm{E}^2}{2!} + \ldots\right) v(t), \\ &= e^{\lambda\mathrm{E}}v(t), \end{aligned}$$

then

$$\begin{aligned} v(t) &= e^{-\lambda\mathrm{E}}u(t), \\ &= \left(1 - \frac{\lambda\mathrm{E}}{1!} + \frac{\lambda^2\mathrm{E}^2}{2!} - \ldots\right) u(t), \\ &= u(t) - \lambda u(t + 1)/1! + \lambda^2 u(t + 2)/2! - \ldots. \end{aligned}$$

Hence we have the following reciprocal transforms—

$$u(t) = \sum_{n=0}^{\infty} \frac{\lambda^n}{n!} v(t + n)$$

and $$v(t) = \sum_{n=0}^{\infty} (-1)^n \frac{\lambda^n}{n!} u(t + n).$$

Examples

1. Let $u(t) = 1/t$, $\lambda = 1$ for $t > 0$.

If $$1/t = \sum_{n=0}^{\infty} v(t + n)/n!,$$

then

$$\begin{aligned} v(t) &= \sum_0^\infty (-1)^n \frac{1}{(t + n)n!} \\ &= \frac{1}{t} - \frac{1}{(t + 1)1!} + \frac{1}{(t + 2)2!} - \frac{1}{(t + 3)3!} + \ldots \\ &= \int_0^1 e^{-x}x^{t-1}\, dx. \end{aligned}$$

Verification

$$v(t+n) = \int_0^1 e^{-x} x^{t+n-1}\, dx;$$

hence
$$\sum_{n=0}^{\infty} v(t+n)/n! = \sum_{n=0}^{\infty} \int_0^1 e^{-x} x^{t+n-1} (n!)^{-1}\, dx$$
$$= \int_0^1 dx \,.\, e^{-x} \,.\, x^{t-1} \sum_{n=0}^{\infty} x^n/n!$$
$$= \int_0^1 dx \,.\, e^{-x} \,.\, x^{t-1} \,.\, e^{x}$$
$$= \int_0^1 dx \,.\, x^{t-1} = 1/t,$$

if all the series are absolutely convergent.

2. If $u(t)$ is a polynomial of degree r in t then we may derive a finite expression for $v(t)$ to satisfy the equation

$$u(t) = \sum_{n=0}^{\infty} \lambda^n v(t+n)/n!$$

For, since

$$u(t) = e^{\lambda E} v(t)$$
$$v(t) = e^{-\lambda E} u(t) = e^{-\lambda} \,.\, e^{-\lambda\Delta} u(t)$$
$$= e^{-\lambda}[1 - \lambda\Delta/1! + \lambda^2\Delta^2/2! - \ldots]u(t)$$
$$= e^{-\lambda}[u(t) - \lambda\Delta u(t)/1! + \lambda^2\Delta^2 u(t)/2! - \ldots]$$

where the series on the right-hand side is finite, terminating with the term $(-1)^r\lambda^r\Delta^r u(t)/r!$. Accordingly $v(t)$ is also a polynomial of degree r in t.

EXERCISES (XIX)

1. If $t^2 = v(t) - v(t+1)/1! + v(t+2)/2! - \ldots$ show that
$$v(t) = (t^2 + 2t + 2)/e.$$

2. If $\sum_0^{\infty} \lambda^n v(t+n)/n! = \sin pt$ show that $v(t) = e^{-\lambda\cos p} \sin(pt - \lambda \sin p)$.

3. If $\sum_0^{\infty} \lambda^n v(t+n) = a^t$ where $a \neq 1/\lambda$, find $v(t)$.

4. $u(t) = v(t) + \lambda v(t+1) + \lambda^2 v(t+2) + \ldots$ and $0 = v(t) + \Delta u(t)$, show that $\lambda = 1$.

5. If $u(t) - \lambda u(t+1)/1! + \lambda^2 u(t+2)/2! - \ldots$
$$= v(t) + \lambda v(t+1)/1! + \lambda^2 v(t+2)/2! + \ldots,$$
show that $u(t) = e^{2\lambda}[v(t) + 2\lambda\Delta v(t)/1! + 4\lambda^2\Delta^2 v(t)/2! + \ldots]$. Examine the case $v(t) = 1/t$.

6. Leibniz' Theorem for Differences

Consider $\Delta[u(x) \,.\, v(x)]$ where $u(x)$ and $v(x)$ are any functions of x. If, as before, E_1 operates only on $u(x)$ and E_2 on $v(x)$, then

$$\Delta u(x) \,.\, v(x) = u(x+1)v(x+1) - u(x)v(x) = (E_1E_2 - 1)u(x)v(x).$$

Since the middle term is itself the difference of products of functions of x

$$\Delta^2 u(x)v(x) = (E_1E_2 - 1)^2 u(x)v(x).$$

Thus, generalizing for r, a positive integer,

$$\begin{aligned}\Delta^r u(x)v(x) &= (E_1E_2 - 1)^r u(x)v(x)\\ &= \Big(E_1{}^r E_2{}^r - rE_1{}^{r-1}E_2{}^{r-1}\\ &\qquad + \frac{r(r-1)}{2!}E_1{}^{r-2}E_2{}^{r-2} - \ldots (-1)^r\Big)u(x)v(x)\\ &= u(x+r)v(x+r) - ru(x+r-1)v(x+r-1)\\ &\qquad + \ldots (-1)^r u(x)v(x),\end{aligned}$$

which is the analogue of Leibniz' theorem.

Alternatively,

$$\Delta = E_1E_2 - 1 = E_1(1 + \Delta_2) - E_1 + \Delta_1 = E_1\Delta_2 + \Delta_1.$$

Hence

$$\begin{aligned}\Delta^r[u(x)v(x)] &= (E_1\Delta_2 + \Delta_1)^r[u(x)v(x)]\\ &= \Big(E_1{}^r\Delta_2{}^r + rE_1{}^{r-1}\Delta_2{}^{r-1}\Delta_1 + \frac{r(r-1)}{2!}E_1{}^{r-2}\Delta_1{}^2\\ &\qquad + \ldots + \Delta_1{}^2\Big)[u(x)v(x)]\\ &= u(x+r)\Delta^r v(x) + r\Delta u(x+r-1)\Delta^{r-1}v(x)\\ &\qquad + \frac{r(r-1)}{2!}\Delta^2 u(x+r-2)\Delta^{r-2}v(x)\\ &\qquad + \ldots + \Delta^r u(x)v(x).\end{aligned}$$

This is a suitable form if either $u(x)$ or $v(x)$ is a polynomial.

Example

$$\begin{aligned}\Delta^r[x^2v(x)] &= (x+r)^2\Delta^r v(x) + r(2x + 2r - 1)\Delta^{r-1}v(x)\\ &\qquad + r(r-1)\Delta^{r-2}v(x).\end{aligned}$$

EXERCISES (XX)

1. Show that the equation
$$(x + r)^2\Delta^2 y(x) + r(2x + 2r - 1)\Delta y(x) + r(r - 1)y(x) = 0,$$
has a solution of the form
$$y(x) = \Delta^{r-2}\left[\frac{\text{polynomial of degree } r - 1}{x^2}\right].$$

2. Using Leibniz' theorem, write down the expression for $\Delta^r(u_n{}^2)$.

3. Express $\Delta^2(u_n v_n)$ entirely in terms of the differences of u_n and v_n.

4. Show that
$$\Delta^r(n^2a^n) = a^n(a - 1)^{r-2}[(a - 1)^2(n + r)^2 + (a - 1)r(2n + 1)(n + r - 1)^2 + r(r - 1)(n + r - 2)^2]$$
where $a \neq 1$.

MISCELLANEOUS EXERCISES

1. Given that
$$\phi(z) = a_0 + a_1 z + a_2 z^2 + \ldots,$$
and
$$1/\phi(z) = b_0 + b_1 z + b_2 z^2 + \ldots,$$
and
$$F(x) = a_0 f(x) + a_1 f(x + 1) + a_2 f(x + 2) + \ldots,$$
show that
$$f(x) = b_0 F(x) + b_1 F(x + 1) + b_2 F(x + 2) + \ldots.$$

2. Show that

(i) $$(x)_r + \frac{(x + 1)_r}{1!} + \frac{(x + 2)_r}{2!} + \ldots = e[(x)_r + r(x)_{r-1} + r(r - 1)(x)_{r-2} + \ldots + r!].$$

(ii) $$\frac{(x + 1)_r}{1!} - \frac{(x + 3)_r}{3!} + \frac{(x + 5)_r}{5!} + \ldots$$
$$= \sin 1\left[(x)_r - \frac{r(r - 1)}{2!}(x)_{r-2} + \frac{r(r - 1)(r - 2)(r - 3)}{4!}(x)_{r-4}\ldots\right] + \cos 1\left[\frac{r}{1!}(x)_{r-1} - \frac{r(r - 1)(r - 2)}{3!}(x)_{r-3} + \ldots\right]$$
the expressions on the right terminating.

3. Show that

(i) $$\Delta^r(x(x - 1)\sin x\pi) = (-2)^{r-2}\sin x\pi[4x(x - 1) - 4rx + r(r - 1)].$$

(ii) $$\Delta^r[x^2 y(x)] = x^2\Delta^r y(x) + r(2x + 1)\Delta^{r-1}y(x + 1). + r(r - 1)\Delta^{r-1}y(x + 2).$$

(iii) $\Delta^2(x^2\tan\pi x) = 2\tan\pi x$.

4. Express $$a_0 + a_1 n + a_2 n^2 + a_3 n^3$$
in the form $b_0 + b_1 n + b_2 n(n - 1) + b_3 n(n - 1)(n - 2)$,
and hence determine $\Delta^{-2}(a_0 + a_1 n + a_2 n^2 + a_3 n^3)$.

5. Sum the series $1 + \dfrac{3x}{1!} + \dfrac{9x^2}{2!} + \dfrac{25x^3}{3!} + \dfrac{57x^4}{4!} + \ldots$.

6. Show that $(\Delta + 2)\sin(\pi x) = 0$.

CHAPTER 2

Interpolation and Extrapolation

FOR a set of values of x, a set of corresponding values of y is given thus—

$$\begin{array}{cccccccc} x_1 & x_2 & x_3 & x_4 & x_5 & x_6 & \dots & x_r \\ y_1 & y_2 & y_3 & y_4 & y_5 & y_6 & \dots & y_r. \end{array}$$

The problems of interpolation and of extrapolation as they are usually presented are concerned with the estimation of y at some position x which is not one of the given set. If

$$x_1 < x_2 < x_3 < \dots < x_r,$$

and if the value of x at which y is required lies in the range $x_1 - x_r$ the act of estimation is called *interpolation*; if y is required at a value of x either greater than x_r or less than x_1, it is called *extrapolation*.

To make the problem more specific suppose the sequence of numbers 1, 3, 5, 7, 9, 11, 13, . . ., 17, 19, 31 be given. It is known that a number exists in the eighth place. In spite of outward appearances there is no logical necessity that compels us to say that the eighth number is 15 or that if the sequence is to be extended generally, the nth number will be $2n - 1$. On the other hand, if this were a sequence of physically measured quantities—say a set of atomic numbers or the distance of a body from a fixed position after unit intervals of time—every experimentalist would conclude, tentatively at least, that the missing number is 15 and the general expression for the number is $2n - 1$. This difference in assumption is one of the features that distinguishes mathematics from an experimental science. When therefore we raise the problem of interpolation or of extrapolation we are concerned with certain physical assumptions and the use of mathematics in the handling of associated data. It will be assumed, for example, that there exists a value of y for every value of x over the range considered. The mathematical procedure thereafter to effect interpolation or extrapolation is merely to seek a functional expression for y which has such a value at every position x, and whose values coincide with those given in the initial set; or if they do not so coincide, deviate from them by less than some specified amount.

Before the mathematician can proceed to find the precise form of this function the given sequence of numbers must be scrutinized in order to discover, if possible, an appropriate *form* of function, one suggested by the *trend* of the given numbers. In the particular case above, for example, it is obvious that the trend is to add 2 to each number to obtain the one which follows. If the mathematician accepts this as an assumption valid everywhere then, of course, $2n - 1$ is inescapable as the nth term. It is, however, by no means the only such assumption that could be made. One might assume, for instance, that the 8th, 16th, 24th . . . numbers were always zero.

The first step therefore is to examine the data for *trend*. Such an examination turns on the nature of the tools the mathematician has at his disposal, and when possible on the physical meaning of the data. If, for example, the set of numbers is presented as successive measurements of the same unchanging thing, the examination would be of a statistical nature and we may pose the question "Is the 'trend' to lie about some central value and to distribute themselves systematically about it?" If, on the other hand, the data represent something which is known to be a function of position, or of time—a basic variable we have called x in this case—then the approach would be made in a rather different way. Here we shall assume that the numbers y can be taken as "accurate" values of a function of x and the mathematical problem is to determine a *suitable* function from which y can be calculated at values of x other than those given. The crux of the question is the meaning to be attached to "suitable." The mathematician has a ready familiarity with certain standard forms of function—polynomial, trigonometrical, exponential, logarithmic, gamma, etc. From his knowledge of the properties of such functions he is able to subject the data to scrutiny and so decide whether they also show trends consistent with those properties. This helps him to select the most suitable form of representation. We shall assume that the values of y are given at equal intervals of x.

Remembering the simple elementary property of a polynomial, viz. differences of the nth order of a polynomial function of the nth degree are constant, it is at once suggested that a table of differences be constructed from the data to discover whether at some stage the differences become constant. If this should indeed turn out to be the case the mathematical problem then resolves itself into determining the precise polynomial from which the interpolated or extrapolated values can be calculated. As we shall see shortly, it is not always necessary to obtain the actual polynomial since the desired values may frequently be calculated directly from the data by a method which, nevertheless, depends on the assumption that the function is a polynomial.

An immediate extension can now be made in our mode of scrutiny of the original data. If the table of differences does not finally yield constant, or at least approximately constant values at some stage, it is possible that some function of the data may do so. For example suppose

$$x = 1, 2, 3, 4, 5, 6, 7.$$

$$y = 1, 1{\cdot}13, 1{\cdot}65, 3{\cdot}08, 7{\cdot}39, 22{\cdot}80, 90{\cdot}00.$$

A table of differences of y leads to

x	y	Δy	$\Delta^2 y$	$\Delta^3 y$
1	1			
		0·13		
2	1·13		0·39	
		0·52		0·52
3	1·65		0·91	
		1·43		1·87
4	3·08		2·88	
		4·31		6·22
5	7·39		9·10	
		15·41		42·69
6	22·80		51·79	
		67·20		
7	90·00			

There is clearly no sign of constancy in the differences. It follows that—

$$y = a_0 + a_1x + a_2x^2 + \ldots + a_nx^n$$

is not a suitable mode of representation of y. If, however, we replace each y by $\log_{10} y$, we have the following table—

x	$\log_{10} y$	$\Delta \log_{10} y$	$\Delta^2 \log_{10} y$
1	0·00000		
		0·05308	
2	0·05308		0·11138
		0·16440	
3	0·21748		0·10667
		0·27107	
4	0·48855		0·10902
		0·38009	
5	0·86864		0·10920
		0·48929	
6	1·35793		0·10602
		0·59631	
7	1·95424		

Since $\Delta^2 \log_{10} y$ is constant (to two decimal places) $\log_{10} y$ should be capable of approximate expression in the form

$$\log_{10} y = a_0 + a_1 x + a_2 x^2,$$

i.e.

$$y = 10^{(a_0 + a_1 x + a_2 x^2)}.$$

If, however, these had not led to a constant column of differences we might try some other function of y. Consider, for example

x	y	Δy	$\Delta^2 y$	$\Delta^3 y$
1	0·020			
		0·673		
2	0·693		− 0·167	
		0·406		− 0·053
3	1·099		− 0·114	
		0·292		− 0·044
4	1·391		− 0·070	
		0·220		
5	1·621			

Again there is no appearance of approach to constancy. We note that the y's increase rather slowly. We try the function e^y, leading to the table

x	e^y	Δe^y	$\Delta^2 e^y$
1	1·020		
		0·980	
2	2·000		0·021
		1·001	
3	3·001		0·017
		1·018	
4	4·019		0·021
		1·039	
5	5·058		

Since the second differences are constant to the second decimal place, we conclude that e^y can be expressed approximately as a quadratic function of x and y in the form

$$y = \log_e (a_0 + a_1x + a_2x^2).$$

Thus the simple principle that the nth difference of a polynomial of the nth degree is constant will enable us to express a tabulated function when suitable in one of the forms

$$y = a_0 + a_1x + \ldots + a_nx^n,$$
$$y = e^{a_0+a_1x+\ldots+a_nx_n}$$
$$y = \log (a_0 + a_1x + \ldots + a_nx^n),$$

and by taking squares or by taking reciprocals of the given data

$$y^2 = a_0 + a_1x + \ldots + a_nx^n,$$
$$y = 1/[a_0 + a_1x + \ldots + a_nx^n].$$

All this then turns on our power to express in convenient polynomial form a set of data whose nth differences are constant.

Fitting the Polynomial

Let us suppose that the function is tabulated at equal intervals h of the independent variable ranging from a to $a + rh$, thus

a	$a + h$	$a + 2h$	$\ldots$	$a + rh$
$f(a)$	$f(a + h)$	$f(a + 2h)$	$\ldots$	$f(a + rh)$.

Thus if we consider $f(a + xh)$ and regard x as the current variable, x increases by unity as we pass from one step to the next. Moreover, as x varies from 0 to 1 the number $a + xh$ varies continuously from a to $a + h$. If we wish to interpolate a value somewhere between $f(a)$ and $f(a + h)$ it will correspond to some value of x between 0 and 1 for $f(a + xh)$. Obviously if we wish we can always arrange to give a such a value as will make the point of interpolation fall between a and $a + h$. This means in fact that we need never concern ourselves with values of x outside the range $0 \leqslant x < 1$. On this understanding we can write

$$\mathrm{E}f(a) = f(a + h), \qquad \mathrm{E}^x f(a) = f(a + xh).$$

Accordingly, from the two identities

$$\mathrm{E}^x = (1 + \Delta)^x$$

and
$$\mathrm{E}^x = \left(\frac{\mathrm{E}}{\mathrm{E} - \Delta}\right)^x = \frac{1}{(1 - \Delta\mathrm{E}^{-1})^x} = (1 - \Delta\mathrm{E}^{-1})^{-x},$$

we obtain

$$\begin{aligned} f(a + xh) &= \mathrm{E}^x f(a) = (1 + \Delta)^x f(a) \\ &= f(a) + x\Delta f(a) + \frac{x(x-1)}{2!}\Delta^2 f(a) \\ &\quad + \ldots + \frac{x(x-1)\ldots(x-r+1)}{r!}\Delta^r f(a), \end{aligned}$$

$$\begin{aligned} f(a + xh) &= \mathrm{E}^x f(a) = (1 - \Delta\mathrm{E}^{-1})^{-x} f(a) \\ &= f(a) + x\Delta f(a - h) + \frac{x(x+1)}{2!}\Delta^2 f(a + 2h) \\ &\quad + \ldots + \frac{x(x+1)\ldots(x+r-1)}{r!}\Delta^r f(a - rh), \end{aligned}$$

the series terminating at the rth difference if the polynomial function is of the rth degree.

These two expressions must, of course, be equivalent. We propose to illustrate this by using both formulae to determine the polynomial corresponding to the following table of values, on p. 42.

Since the second differences are constant we expect the function to be of the second degree. Taking $a = 0$ and $h = 1$, and using the first formula,

$$f(x) = 0 + x\,.\,(0) + \frac{x(x-1)}{2!}\,.\,(2) = x^2 - x.$$

This corresponds to following the downward direction of the arrows. Using the second formula,

$$f(x) = 0 + x(-2) + \frac{x(x+1)}{2!} \cdot (2) = x^2 - x.$$

This corresponds to following the upward direction of the double arrows. Both lead to the same expression, x being increased from the zero of $a + xh$.

$a + xh$	$f(a + xh)$	Δ	Δ^2	Δ^3
-5	30			
		-10		
-4	20		2	
		-8		0
-3	12		2	
		-6		0
-2	6		2	
		-4		0
-1	2		2	
		-2		0
0	0		2	
		0		0
1	0		2	
		2		0
2	2		2	
		4		0
3	6		2	
		6		
4	12			

The first of these interpolation formulae depends on the values of the function on the positive side of x only; the second on the values to the negative side of x. Thus the first would be used for interpolation when $x = 0$ is the lower bound of the tabulated interval, the second when it is the upper bound. This is, of course, not essential here, but if we depart from this scheme we would in fact be extrapolating instead of interpolating. Again this would be of no importance if we knew that the function tabulated were actually a polynomial. In fact, however, the method is used for *any* function on the assumption that over the region tabulated and for which interpolation is required the function can be adequately represented by a polynomial. This assumption is justified in practice if we find that the tabulated values produce approximately zero difference at a certain stage.

Example

Interpolation of the value of $J_0(z)$ at $z = 5{\cdot}225$, given the values of $J_0(z)$ at $z = 4{\cdot}4, 4{\cdot}6, 4{\cdot}8, \ldots, 5{\cdot}8, 6{\cdot}0$.

x	z	$J_0(z)$	$\Delta J_0(z)$	$\Delta^2 J_0(z)$	$\Delta^3 J_0(z)$
− 4	4·4	− 0·34226			
			0·04612		
− 3	4·6	− 0·29614		0·00959	
			0·05571		− 0·00247
− 2	4·8	− 0·24043		0·00712	
			0·06283		− 0·00264
− 1	5·0	− 0·17760		0·00448	
			0·06731		− 0·00271
0	5·2	− 0·11029		0·00177	
			0·06908		− 0·00267
1	5·4	− 0·04121		− 0·00090	
			0·06818		− 0·00255
2	5·6	+ 0·02697		− 0·00345	
			0·06473		− 0·00294
3	5·8	+ 0·09170		− 0·00639	
			0·05834		
4	6·0	+ 0·15064			

In this case we take $z = 5{\cdot}2 + 0{\cdot}2x$, i.e. $a = 5{\cdot}2$ and $h = 0{\cdot}2$. Thus, when $z = 5{\cdot}225$, $x = 0{\cdot}125$. The third differences of the function are constant to three decimal places which indicates that over this range the function may be represented by a cubic in x.

Using the first of the two interpolation formulae, viz.—

$$f(a + hx) = f(a) + x\Delta f(a) + \ldots + \frac{x(x-1)\ldots(x-r+1)}{r!}\Delta^r f(a)$$

we obtain,

$$\begin{aligned} J_0(5{\cdot}225) &= -0{\cdot}11029 + 0{\cdot}008635 + 0{\cdot}000049 - 0{\cdot}000086 \\ &= -0{\cdot}10169. \end{aligned}$$

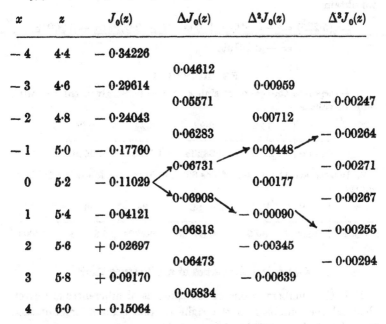

The tabulated value of $J_0(5{\cdot}225)$ is $-0{\cdot}10170$, thus the above calculated figure is correct to four decimal places. Using the formula—

$$f(a + xh) = f(a) + x\Delta f(a - h) + \frac{x(x+1)}{2!}\Delta^2 f(a - 2h) + \dots + \frac{x(x+1)\dots(x+r-1)}{r!}\Delta^r f(a - rh),$$

we obtain,

$$J_0(5{\cdot}225) = -0{\cdot}11029 + 0{\cdot}00841 + 0{\cdot}000315 - 0{\cdot}00013$$
$$= -0{\cdot}10169.$$

EXERCISES (I)

1. Interpolate the value of sinh x at $x = 0{\cdot}232$ given the following table of values—

x	0·22	0·23	0·24	0·25
sinh x	0·22178	0·23203	0·24231	0·25261

2. Interpolate the value of cos 30° 18′ from the following table of values—

$x°$	28	30	32	34	36	38
cos x	0·88295	0·86603	0·84805	0·82904	0·80902	0·78801

Elementary Properties of a Difference Table

1. If the numbers in one of the columns of differences are given then all the numbers to the right of this column are obtained immediately by successive differencings.

2. If all the numbers in one of the columns of differences and one number in the preceding column are given, then all the numbers in the preceding column are immediately calculable.

3. If one number in each column is given and if the $(r + 1)$th column of differences is zero (so that the polynomial function representing the table of values is of the rth degree) then the whole table is constructible and the polynomial is itself determinate.

These three elementary properties should be obvious from the manner in which the table is constructed.

4. It follows from this that if any broken line be drawn across the table from any number in the first column to any number in the

second, then to any number in the third, and so on until the column of zeros is reached, then the coefficients of the polynomial must be equally expressible in terms of these successive numbers.

5. All routes across the table lead to the same polynomial. The properties 4 and 5 are illustrated by the two expressions already derived for $f(a + xh)$ where the one expression is associated with the route that slopes upwards and the other with the route that slopes downwards.

6. We may replace the property under 3 above by the following statement—If a single relation between the numbers in each column be given, and if the $(r + 1)$th column of differences is zero (so that the polynomial function representing the table of values is of the rth degree) then the whole table is constructible and the polynomial is itself determinate.

From these elementary considerations it should now be obvious that instead of there being merely two ways of expressing the function $f(a + xh)$ that we have found, there must be as many ways as there are routes across the table, plus those that arise from the extra latitude given to us by the fact involved in 6. We shall see shortly how this is to be used. Moreover we have seen that the two forms already found for $f(a + xh)$ point to different sets of circumstances in which they can be most suitably used. We shall therefore now develop other forms for $f(a + xh)$ corresponding to other routes across the table and indicate in what circumstances they also will be suitable for interpolation.

Consider the following table—

$f(a - 3h)$			
	$\Delta f(a - 3h)$		
$f(a - 2h)$		$\Delta^2 f(a - 3h)$	
	$\Delta f(a - 2h)$		$\Delta^3 f(a - 3h)$
$f(a - h)$		$\Delta^2 f(a - 2h)$	
	$\Delta f(a - h)$		$\Delta^3 f(a - 2h)$
$f(a)$		$\Delta^2 f(a - h)$	
	$\Delta f(a)$		$\Delta^3 f(a - h)$
$f(a + h)$		$\Delta^2 f(a)$	
	$\Delta f(a + h)$		$\Delta^3 f(a)$
$f(a + 2h)$		$\Delta^2 f(a + h)$	
	$\Delta f(a + 2h)$		
$f(a + 3h)$			

It will be more convenient to write this table symbolically where every operator is regarded as acting on $f(a)$.

E^{-3}			
	ΔE^{-3}		
E^{-2}		$\Delta^2 E^{-3}$	
	ΔE^{-2}		$\Delta^3 E^{-3}$
E^{-1}		$\Delta^2 E^{-2}$	
	ΔE^{-1}		$\Delta^3 E^{-2}$
1		$\Delta^2 E^{-1}$	
	Δ		$\Delta^3 E^{-1}$
E		Δ^2	
	ΔE		Δ^3
E^2		$\Delta^2 E$	
	ΔE^2		$\Delta^3 E$
E^3		$\Delta^2 E^2$	
	ΔE^3		
E^4			

Since $f(a + xh) = E^x f(a)$ our problem is to express E^x in terms of the elements that lie at the ends of each broken line in the total route across the table. For example, in the first of the two formulae already derived E^x was expressed in terms of $1, \Delta, \Delta^2, \Delta^3 \ldots$, i.e. $E^x = A_0 + A_1\Delta + A_2\Delta^2 + \ldots$, the coefficients A_0, A_1, A_2 being, of course, functions of x. This was found in effect by writing $E^x = (1 + \Delta)^x$ and expanding, in which case we found

$$A_0 = 1, \qquad A_1 = x, \qquad A_2 = x(x-1)/2!, \qquad \ldots.$$

In the second of these formulae E^x was expressed in terms of

$$1, \quad \Delta E^{-1}, \quad \Delta^2 E^{-2}, \quad \Delta^3 E^{-3}, \ldots,$$

i.e. $$E^x = A_0 + A_1(\Delta E^{-1}) + A_2(\Delta E^{-1})^2 + A_3(\Delta E^{-1})^3 + \ldots,$$

a power series in ΔE^{-1}.

Now $$\Delta E^{-1} = (E - 1)E^{-1} = 1 - E^{-1},$$

then $$E = (1 - \Delta E^{-1})^{-1},$$

so that the required formula was found by expanding

$$E^x = (1 - \Delta E^{-1})^{-x}$$

in ascending powers of ΔE^{-1}.

Now consider the expression for $f(a + xh)$ that would be obtained by following the zig-zag route specified in succeeding columns by the symbolical elements $1, \Delta, \Delta^2 E^{-1}, \Delta^3 E^{-1}, \Delta^4 E^{-2}, \Delta^5 E^{-2}, \ldots$.

Thus we have to express E^x in the following form—

$$E^x = A_0 + A_2\Delta^2E^{-1} + A_4\Delta^4E^{-2} + \ldots + A_{2r}\Delta^{2r}E^{-r} + \ldots$$
$$+ A_1\Delta + A_3\Delta^3E^{-1} + A_5\Delta^5E^{-2} + \ldots + A_{2r+1}\Delta^{2r+1}E^{-r} + \ldots.$$

If E is everywhere replaced by $1 + \Delta$, every power of E expanded in powers of Δ, and the first few coefficients of powers of Δ equated on both sides we easily find

$$A_0 = 1,$$
$$A_1 = x,$$
$$A_2 = (x-1)x/2\,!,$$
$$A_3 = (x-1)x(x+1)/3\,!,$$
$$A_4 = (x-2)(x-1)x(x+1)/4\,!.$$

To obtain the general term it is more convenient to operate on both sides of the general expression for E^x above with E^r, then replacing E by $1 + \Delta$ it becomes—

$$(1+\Delta)^{x+r} = A_0(1+\Delta)^r + A_2\Delta^2(1+\Delta)^{r-1} + \ldots + A_{2r}\Delta^{2r}$$
$$+ A_{2r+2}\Delta^{2r+2}(1+\Delta)^{-1} + \ldots + A_1\Delta(1+\Delta)^r$$
$$+ \ldots + A_{2r-1}\Delta^{2r-1}(1+\Delta) + A_{2r+1}\Delta^{2r+1} + \ldots.$$

Now equating coefficients of Δ^{2r+1} and Δ^{2r+2} on both sides we notice that on the right-hand side these terms each arise at only one place, viz. in association with A_{2r+1} and A_{2r+2} respectively. On the left-hand sides the coefficients of these powers of Δ are immediately written down from the binomial expansion. Accordingly we find

$$A_{2r+1} = (x+r-1)\ldots(x-r)/(2r+1)\,!$$
$$A_{2r+2} = (x+r)(x+r-1)\ldots(x-r-1)/(2r+2)\,!$$

With these coefficients therefore we use the original operator E^x on $f(a)$ and derive finally—

$$f(a+xh) = f(a) + \frac{(x-1)x}{2\,!}\Delta^2 f(a-h) + \ldots$$
$$+ \frac{(x-r)(x-r+1)\ldots(x+r-1)}{2r\,!}\Delta^{2r}f(a-rh) + \ldots$$
$$+ x\Delta f(a) + \frac{(x-1)x(x+1)}{3\,!}\Delta^3 f(a-h) + \ldots$$
$$+ \frac{[x-(r+1)](x-r)\ldots(x+r)}{(2r+1)\,!}\Delta^{2r+1}f(a-rh) + \ldots,$$

the series finally terminating when the order of difference that is zero is reached.

In what circumstances is this form of the polynomial suitable for interpolation purposes? We notice that the first term is $f(a)$ and therefore if the interpolation is in the neighbourhood of a on the positive side, the principal part of the interpolated value can be expected to be $f(a)$. The next part is clearly $x\Delta f(a)$ which brings in roughly the proportionate effect of the change in value for $f(a)$ to $f(a+h)$. The next term $\Delta^2 f(a-h)$ depends now also on terms on the other side of a and so, step-by-step, the values on both sides of the interpolated position are brought to bear to add their quota in determining the interpolated value. This formula therefore is suitable for the case where the interpolation is close to a, and where the given values are ranged on both sides of the interpolated position. In the calculation the terms $f(a)$, $x\Delta f(a)$, $\dfrac{(x-1)x}{2!}\Delta^2 f(a-h)$, may be expected to contribute successively small amounts towards the final value of $f(a+xh)$ so that in a sense the order of accuracy of the interpolation may be gauged as the calculation proceeds.

The formula for interpolation just derived is clearly not very convenient for the case where the point of interpolation lies close to the mid-point of a range, for in that case the value required would be approximately the mean of the values $f(a)$ and $f(a+h)$. In fact on the assumption that $f(a+xh)$ is approximately the mean of $f(a)$ and $f(a+h)$ we can easily evaluate the next term in this way—

$$\begin{aligned}\text{Error} &= f(a+xh) - \tfrac{1}{2}[f(a) + f(a+h)] \\ &= [E^x - \tfrac{1}{2}(1+E)]f(a) = [(1+\Delta)^x - 1 - \tfrac{1}{2}\Delta]f(a) \\ &= (x-\tfrac{1}{2})\Delta f(a) + \frac{x(x-1)}{2!}\Delta^2 f(a) + \ldots,\end{aligned}$$

so that *approximately*,

$$f(a+xh) = \tfrac{1}{2}[f(a) + f(a+h)] + (x-\tfrac{1}{2})\Delta f(a)$$

and clearly when x is close to $\tfrac{1}{2}$ the second term is correspondingly small.

To develop a formula which will bring these considerations into effect we trace the route indicated in the following symbolical table where the vertical arrows imply that the mean values of the terms so linked are to be taken as terminal points of the broken line.

$$
\begin{array}{ccccccccc}
E^{-4} & & & & & & & & \\
& \Delta E^{-4} & & & & & & & \\
E^{-3} & & \Delta^2E^{-4} & & & & & & \\
& \Delta E^{-3} & & \Delta^3E^{-4} & & & & & \\
E^{-2} & & \Delta^2E^{-3} & & \Delta^4E^{-4} & & & & \\
& \Delta E^{-2} & & \Delta^3E^{-3} & & & & & \\
E^{-1} & & \Delta^2E^{-2} & & \Delta^4E^{-3} & & & & \\
& \Delta E^{-1} & & \Delta^3E^{-2} & & & & & \\
1 & & \Delta^2E^{-1} & & \Delta^4E^{-2} & & & & \\
\updownarrow\longrightarrow & \Delta\longrightarrow & \updownarrow\longrightarrow & \Delta^3E^{-1}\longrightarrow & \updownarrow\longrightarrow & & & & \\
E & & \Delta^2 & & \Delta^4E^{-1} & & & & \\
& \Delta E & & \Delta^3 & & & & & \\
E^2 & & \Delta^2E & & \Delta^4 & & & & \\
& \Delta E^2 & & \Delta^3E & & & & & \\
E^3 & & \Delta^2E^2 & & & & & & \\
& \Delta E^3 & & & & & & & \\
E^4 & & & & & & & &
\end{array}
$$

It may be seen that as nearly as can be achieved the numbers brought into the calculation lie symmetrically about the position of interpolation. In this case we have to find the symbolical relationship of the form

$$
\begin{aligned}
E^x &= A_0(1 + E)/2 + A_1\Delta + A_2(\Delta^2E^{-1} + \Delta^2)/2 + A_3\Delta^3E^{-1} \\
&\qquad + A_4(\Delta^4E^{-2} + \Delta^4E^{-1})/2 + \ldots \\
&= \tfrac{1}{2}[A_0 + A_2\Delta^2E^{-1} + A_4\Delta^4E^{-2} + \ldots + A_{2r}\Delta^{2r}E^{-r} + \ldots](1 + E) \\
&\qquad + [A_1\Delta + A_3\Delta^3E^{-1} + \ldots + A_{2r+1}\Delta^{2r+1}E^{-r} + \ldots].
\end{aligned}
$$

By replacing E by $1 + \Delta$ and expanding as far as Δ^3 on both sides and equating powers of Δ we immediately obtain,

$$
\begin{aligned}
A_0 &= 1, \\
A_1 &= x - \tfrac{1}{2}, \\
A_2 &= (x - 1)x/2!, \\
A_3 &= (x - 1)x(x - \tfrac{1}{2})/3!
\end{aligned}
$$

This, however, does not lead so easily to the general term. Accordingly, as in the previous case we operate on both sides with E^r and replace E by $1 + \Delta$,

$$
\begin{aligned}
(1 + \Delta)^{x+r} = \tfrac{1}{2}[(1 + \Delta)^r + A_2\Delta^2(1 + \Delta)^{r-1} + \ldots + A_{2r}\Delta^{2r} \\
+ A_{2r+2}\Delta^{2r+2}(1 + \Delta)^{-1} + \ldots](2 + \Delta) \\
+ [A_1\Delta(1 + \Delta^r) + \ldots + A_{2r+1}\Delta^{2r+1} + \ldots].
\end{aligned}
$$

Equating the coefficients of Δ^{2r+1} and Δ^{2r+2} on both sides we obtain—

$$\tfrac{1}{2}A_{2r} + A_{2r+1} = \frac{(x+r)(x+r-1)\ldots(x-r)}{(2r+1)!},$$

$$A_{2r+2} = \frac{(x+r)(x+r-1)\ldots(x-r-1)}{(2r+2)!}.$$

By replacing r by $r-1$ in the second of these we obtain

$$A_{2r} = \frac{(x+r-1)(x+r-2)\ldots(x-r)}{(2r)!},$$

and inserting this in the first of these

$$A_{2r+1} = \frac{(x+r-1)(x+r-2)\ldots(x-r)}{(2r+1)!}(x-\tfrac{1}{2}).$$

This therefore determines the coefficients in the polynomial. For purposes of calculation it is clear that

$$A_{2r+1} = \frac{x-\frac{1}{2}}{2r+1}A_{2r}, \qquad A_{2r+2} = \frac{(x+r)(x-r-1)}{(2r+2)(2r+1)}A_{2r},$$

so that starting with

$$A_0 = 1, \quad A_1 = x - \tfrac{1}{2}, \quad A_2 = \frac{x(x-1)}{2!},$$

we easily derive $A_3, A_4 \ldots$ in succession. All that is required therefore is to build up the appropriate table of differences, to separate out the items on the route with the appropriate differences and their means on that route, and to multiply each by its coefficient calculated step-by-step each from the preceding one.

Example

Given the following values for $J_1(z)$, interpolate the value of $J_1(z)$ at $z = 1{\cdot}053$.

z	0·800	0·900	1·000	1·100	1·200	1·300
$J_1(z)$	0·368842	0·405950	0·440051	0·470902	0·498289	0·522023

For convenience we make the following definitions—

$$1 + x/10 = z$$

$$f(x) = 100J_1(z) = 100J_1(1 + x/10).$$

Thus we have the following difference table

x	$f(x)$	$\Delta f(x)$	$\Delta^2 f(x)$	$\Delta^3 f(x)$	$\Delta^4 f(x)$
− 2	36·8842				
		3·7108			
− 1	40·5950		− 0·3007		
		3·4101		− 0·0243	
0	44·0051		− 0·3250		0·0029
	↕ →	3·0851 →	↕ →	− 0·0214 →	↕ →
1	47·0902		− 0·3464		0·0025
		2·7387		− 0·0189	
2	49·8289		− 0·3653		
		2·3734			
3	52·2023				

If $z = 1·053$, then $x = 0·53$, and $f(0·53) = 100J_1(1·053)$. Using the interpolation formula established above and following the path marked on the table we obtain

$$f(0·53) = 45·547650 + A_1 3·0851 - A_2 0·3357 - A_3 0·0214 + A_4 0·0027.$$

Now
$$A_1 = x - \tfrac{1}{2} = 0·03,$$
$$A_2 = x(x-1)/2 = -0·12455,$$
$$A_3 = (x - \tfrac{1}{2})A_2/3 = -0·0012455,$$
$$A_4 = \frac{(x+1)(x-2)}{12} A_2 = 0·0233438.$$

Hence
$$f(0·53) = 45·6821.$$
$$J_1(1·053) = 0·456821.$$

(The actual value of $J_1(1·053)$ is 0·456821100, correct to nine decimal places.)

Applying these methods the following four formulae for interpolation can be established—

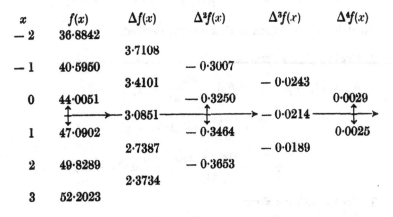

1. Newton–Bessel Formula

$$f(a + \tfrac{1}{2}h + xh) = \tfrac{1}{2}[f(a + h)] + x\Delta f(a)$$
$$+ \frac{x^2 - 1/4}{2!}\,\tfrac{1}{2}[\Delta^2 f(a - h) + \Delta^2 f(a)] + \frac{x(x^2 - 1/4)}{3!}\,\Delta^3 f(a - h)$$
$$+ \frac{(x^2 - 1/4)(x^2 - 9/4)}{4!}\,\tfrac{1}{2}[\Delta^4 f(a - 2h) + \Delta^4 f(a - h)] + \ldots$$

which is simply a form of that already derived.

2. Newton–Gauss Formula

$$f(a + xh) = f(a) + x\Delta f(a) + \frac{x(x - 1)}{2!}\,\Delta^2 f(a - h)$$
$$+ \frac{(x + 1)x(x - 1)}{3!}\,\Delta^3 f(a - h)$$
$$+ \frac{(x + 1)x(x - 1)(x - 2)}{4!}\,\Delta^4 f(a - 2h) + \ldots.$$

3. Newton–Stirling Formula

$$f(a + xh) = f(a) + x\,\frac{\Delta f(a) + \Delta f(a - h)}{2} + \frac{x^2}{2!}\,\Delta^2 f(a - h)$$
$$+ \frac{x(x^2 - 1^2)}{3!}\,\frac{\Delta^3 f(a - h) + \Delta^3 f(a - 2h)}{2} + \frac{x^4}{4!}\,(x^2 - 1^2)\Delta^4 f(a - 2h)$$
$$+ \frac{x(x^2 - 1^2)(x^2 - 2^2)}{2!}\,\frac{\Delta^5 f(a - 2h) + \Delta^5 f(a - 3h)}{2} + \ldots.$$

EXERCISES (II)

1. Show that the polynomial of $(r - 1)$th degree which passes through the r points $(x_1, y_1), (x_2, y_2) \ldots (x_r, y_r)$ is

$$y = A_1(x - x_2)(x - x_3) \ldots (x - x_r)$$
$$+ A_2(x - x_1)(x - x_1) \ldots (x - x_r)$$
$$+ \ldots + A_r(x - x_1)(x - x_2) \ldots (x - x_{r-1}),$$

where
$$A_1 = y_1/(x_1 - x_2)(x_1 - x_3) \ldots (x_1 - x_r),$$
$$A_2 = y_2(x_2 - x_1)(x_2 - x_3) \ldots (x_2 - x_r),$$
$$\ldots\ldots\ldots\ldots\ldots\ldots\ldots\ldots$$
$$A_r = y_r(x_r - x_1)(x_r - x_2) \ldots (x_r - x_{r-1}).$$

[Lagrange]

2. Show that the polynomial that passes through the four points (0, 0), (1, 0), (2, 4), (3, 18) passes also through the points (4, 48), (5, 100), (6, 180).

We have already explained that the differences between these interpolation polynomials reside only in the order in which the terms of one and the same polynomial are arranged for the purpose of arriving quickly at the interpolated values required. The polynomial itself is mathematically unique. In the cases we have developed we have merely illustrated how certain routes across the table are for certain purposes to be preferred to others. As is to be expected there are many such forms dependent on the object in view and on the circumstances.

There is, however, one limitation in all these cases imposed from the beginning. To assume that the function under consideration can be expressed in the form

$$y = a_0 + a_1x + a_2x^2 + \ldots + a_nx^n$$

is to assume that y is single valued for each value of x. This is to exclude a function whose x, y graph doubles back on itself through a point at which $\frac{dy}{dx}$ is infinite—or at the best it is to assume that in the close neighbourhood of such a curve there is an approximate representation in terms of a polynomial—and this definitely limits the use of the polynomial in that case for purposes of extrapolation. Moreover, the formulae as derived assume that the dependent variable has been tabulated at equal intervals of the independent variable although there are modifications to these formulae which enable this difficulty to be overcome. Both these obstacles, however, may frequently be surmounted if we will regard the functional relation between x and y as expressed in parametric form

$$x = f(t), \qquad y = g(t).$$

When corresponding pairs of values of x and y are given even if unevenly spaced, we may tentatively associate them with a variable t which we assume is evenly spaced and then proceed to express x and y separately as polynomials in t. The method is best seen through an illustration.

Example

Given the following pairs of values of (x, y), to find y at $x = 2$.

x	4	6	6	4	0	-6	-14	-24
y	0	5	8	9	8	5	0	-7

Constructing the difference table in x, we have

t	x	Δx	$\Delta^2 x$
0	4		
		+ 2	
1	6		− 2
		0	
2	6		− 2
		− 2	
3	4		− 2
		− 4	
4	0		− 2
		− 6	
5	− 6		− 2
		− 8	
6	− 14		− 2
		− 10	
7	− 24		

Thus, assuming x is a function of a new variable t,

$$\Delta^2 x = -2.$$

Hence

$$x = -t^2 + At + B$$

where A and B are constants. Substituting $t = 0$, $x = 4$ and $t = 1$, $x = 6$ we obtain

$$x = -t^2 + 3t + 4.$$

Similarly

t	y	Δy	$\Delta^2 y$
0	0		
		5	
1	5		− 2
		3	
2	8		− 2
		1	
3	9		− 2
		− 1	
4	8		− 2
		− 3	
5	5		− 2
		− 5	
6	0		− 2
		− 7	
7	− 7		

Thus $\Delta^2 y = -2.$

Hence

$$y = -t^2 + at + b.$$

Using the fact that $t = 0$, $y = 0$ and $t = 1$, $y = 5$, we obtain

$$y = -t^2 + 6t.$$

We interpolate for y when $x = 2$.

If $x = 2 \qquad 2 = -t^2 + 3t + 4;$

hence $t = \dfrac{3 \pm \sqrt{17}}{2} = 3{\cdot}5615 \quad \text{or} \quad -0{\cdot}5615.$

When $t = 3{\cdot}5615 \qquad y = 8{\cdot}6847,$

$t = -0{\cdot}5615 \qquad y = -3{\cdot}6843,$

the former being an interpolated and the latter an extrapolated value.

EXERCISE (III)

Interpolate for y when $x = 1{\cdot}31000$ given that

x	1·22140	1·24608	1·27125	1·29693	1·32313	1·34986	1·37713
y	1·02007	1·02430	1·02894	1·03399	1·03946	1·04534	1·05164

Expression for Differential Coefficients and for Definite Integrals

Once more assuming that it is possible to express a given tabulated function as a polynomial the differential coefficients may be derived directly for any position. For example, if we take the form

$$f(a + xh) = f(a) + x\Delta f(a) + \frac{x(x-1)}{2!}\Delta^2 f(a) + \frac{x(x-1)(x-2)}{3!}\Delta^3 f(a) + \ldots,$$

then

$$hf'(a + xh) = \Delta f(a) + \frac{2x-1}{2!}\Delta^2 f(a) + \frac{3x^2 - 6x + 2}{3!}\Delta^3 f(a) + \ldots$$

$$h^2 f''(a + xh) = \Delta^2 f(a) + (x-1)\Delta^3 f(a) + \ldots$$

and in particular when $x = 0$,

$$hf'(a) = \Delta f(a) - \tfrac{1}{2}\Delta^2 f(a) + \tfrac{1}{3}\Delta^3 f(a) - \ldots,$$

$$h^2 f''(a) = \Delta^2 f(a) - \Delta^3 f(a) + \ldots.$$

Regarded from the operational standpoint, the first of these states that

$$hD = \Delta - \tfrac{1}{2}\Delta^2 + \tfrac{1}{3}\Delta^3 \ldots,$$

and the second is derived merely by squaring both sides of this operational equality, viz.

$$h^2D^2 = \Delta^2 - \Delta^3 + \tfrac{11}{12}\Delta^4 + \ldots.$$

This operational equality is interesting in that it brings out the extent of the field within which algebraic transformations may be utilized in relation to operations. For if we remember that $\log(1 + x)$ is the name for the series

$$x - \tfrac{1}{2}x^2 + \tfrac{1}{3}x^3 - \ldots$$

then on a similar basis we may take $\log(1 + \Delta)$ to be the name for the operation

$$\Delta - \tfrac{1}{2}\Delta^2 + \tfrac{1}{3}\Delta^3 - \ldots,$$

i.e. the differential-difference relation we have found states that

$$hD = \log(1 + \Delta) = \log E.$$

Now, of course, Taylor's theorem states that

$$f(a + h) = f(a) + hf'(a) + \frac{h^2}{2!}f''(a) + \ldots,$$

or, written operationally, the operators acting on $f(a)$,

$$E = 1 + hD + \frac{h^2D^2}{2!} + \frac{h^3D^3}{3!} + \ldots.$$

Remembering that e^x is a short way of writing

$$1 + x + \frac{x^2}{2!} + \frac{x^3}{3!} + \ldots$$

we can say that

$$E = e^{hD}.$$

Accordingly we thus have the two operational equalities

$$hD = \log E \qquad \text{and} \qquad E = e^{hD},$$

showing in fact that the usual relations between the logarithmic and the exponential functions are valid for these operators. So to obtain any order of differentiation, say, $f^{(r)}(a)$, we merely write

$$h^r f^{(r)}(a) = h^r D^r f(a) = (hD)^r f(a)$$
$$= \left(\Delta - \frac{\Delta^2}{2} + \frac{\Delta^3}{3} - \ldots\right)^r f(a),$$

and expand the operator on the right-hand side. Since we are dealing with polynomials all these series of operations are of finite extent.

The Error in the Use of an Interpolation Formula

If in place of any function $f(x)$ we use the polynomial

$$F(x) \equiv f(0) + x\Delta f(0) + \frac{x(x-1)}{2!}\Delta^2 f(0) + \ldots + \frac{x(x-1)\ldots(x-n+1)}{n!}\Delta^n f(0),$$

is it possible to estimate the error committed? Let

$$f(x) = F(x) + \frac{x(x-1)\ldots(x-n)}{(n+1)!}G(x).$$

Then $f(x) = F(x)$ at all integral values of x from 0 to n, and the error at any position x is $f(x) - F(x)$.

Consider the function of ξ and x given by

$$f(\xi) - F(\xi) - \frac{\xi(\xi-1)\ldots(\xi-n)}{(n+1)!}G(x),$$

where $x \neq 0, 1, 2, \ldots, n$, and ξ is regarded as the variable. Then this function vanishes for $\xi = 0, 1, 2, \ldots, n$ and x. Hence its $(n+1)$th derivative with respect to ξ must vanish at some position θ in the whole interval. But the $(n+1)$th derivative with respect to ξ is

$$f^{(n+1)}(\xi) - G(x).$$

Thus
$$G(x) = f^{(n+1)}(\theta).$$

Accordingly

$$f(x) = f(0) + x\Delta f(0) + \frac{x(x-1)}{2!}\Delta^2 f(0) + \ldots + \frac{x(x-1)\ldots(x-n+1)}{n!}\Delta^n f(0) + \frac{x(x-1)\ldots(x-n)}{(n+1)!}f^{(n+1)}(\theta),$$

where $f(x)$ is no longer restricted to be a polynomial. When $f^{(n+1)}(\theta)$ can be shown to lie within finite limits the interpolation formula now enables the interpolation to be effected subject to an error whose greatest value can evidently be estimated.

Approximate Expressions for Integrals

We are now in a position to write down an expression for the indefinite integral of the function. For example, if we write $a = 1$, $h = 1$ so that

$$f(x) = f(0) + x\Delta f(0) + \frac{x(x-1)}{2!} \cdot \Delta^2 f(0) + \ldots,$$

then $$\int_0^x f(x)dx = xf(0) + \frac{x^2}{2}\Delta f(0) + \frac{2x^3 - 3x^2}{12}\Delta^2 f(0) + \ldots,$$

with a corresponding expression derived from each of the interpolation forms. If in the above we write $x = n$ and insert the remainder previously found, then

$$\int_0^n f(x)\,dx = nf(0) + \frac{n^2}{2}\Delta f(0) + \frac{2n^3 - 3n^2}{12}\Delta^2 f(0) + \ldots$$
$$+ \frac{f^{n+1}(\theta)}{(n+1)!}\int_0^n x(x-1)\ldots(x-n)\,dx.$$

EXERCISE (IV)

Show that for an interval h

$$\int_0^{2h} f(x)\,dx = \tfrac{1}{3}h[f(0) + 4f(h) + f(2h)] - \frac{h^5}{90}f^{iv}(2\theta h).$$

Deduce Simpson's Rule, viz.

$$\int_0^{2nh} f(x)\,dx = \tfrac{1}{3}h\{f(0) + 4f(h) + 2f(2h)$$
$$+ 4f(3h) + 2f(4h)$$
$$\ldots\ldots\ldots\ldots$$
$$\ldots\ldots\ldots\ldots$$
$$+ 4f[(2n-1)h] + f(2nh)\} - \frac{nh^5}{90}f^{iv}(2\theta h).$$

where $0 < \theta < 1$.

Symbolic Expression for the Definite Integral

To carry through the operations implied by the expression $\int_a^{a+rh} f(t)\,dt$ we may detail the successive steps as follows—

First we integrate $f(t)$, the indefinite integral being represented by $D^{-1}f(t)$. We then insert the value of t at the upper limit $a + rh$ and subtract from the result the number obtained by inserting the value of t at the lower limit. In fact if

$$\int f(t)\,dt = F(t)$$

then
$$\int_a^{a+rh} f(t)\,dt = F(a + rh) - F(a) = (E^r - 1)F(a)$$
$$= (E^r - 1)\int f(a)\,da = (E^r - 1)D^{-1}f(a).$$

It should now be clear that various expressions for the definite integral on the left will be obtained by using various modes of expansion for the symbolical expression on the right, viz. $(E^r - 1)/D$, the type of expansion depending on whether, in the circumstances, the resulting terms are capable of direct evaluation and whether the convergence is rapid.

Euler–Maclaurin Integral-Summation Formula

We are concerned with the evaluation of

$$\int_a^{a+rh} f(t)\,dt = \frac{(E^r - 1)f(t)}{D}$$

where $Ef(a) = f(a + h)$, the integral being itself a function of a which occurs in the limits. We begin by extracting from the integral its approximate expression through the well-known trapezium rule. Thus let

$$\int_a^{a+rh} f(t)\,dt = h[\tfrac{1}{2}f(a) + f(a + h) + \ldots + f[a + \overline{r - 1}h]$$
$$+ \tfrac{1}{2}f(a + rh)] + R$$
$$= h[\tfrac{1}{2} + E + E^2 + \ldots + E^{r-1} + \tfrac{1}{2}E^r]f(a) + R$$
$$= h[1 + E + E^2 + \ldots + E^{r-1}]f(a)$$
$$- \frac{h}{2}(1 - E^r)f(a) + R$$

$$= h\left(\frac{1 - \mathrm{E}^r}{1 - \mathrm{E}}\right) f(a) - \frac{h}{2}(1 - \mathrm{E}^r) f(a) + R$$

$$= h(\mathrm{E}^r - 1)\left(\frac{1}{\mathrm{E} - 1} + \frac{1}{2}\right) f(a) + R.$$

Accordingly

$$R = h(\mathrm{E}^r - 1)\left(\frac{1}{h\mathrm{D}} - \frac{1}{\mathrm{E} - 1} - \frac{1}{2}\right) f(a).$$

Now
$$1 + \Delta = \mathrm{E} = e^{h\mathrm{D}}$$

or
$$\frac{1}{\Delta} = \frac{1}{e^{h\mathrm{D}} - 1}.$$

The numbers B_0, B_1, B_2 are defined by the relation

$$\frac{z}{e^z - 1} = B_0 + \frac{B_1 z}{1!} + \frac{B_2 z^2}{2!} + \frac{B_3 z^3}{3!} + \ldots .$$

It is easily seen that the coefficients B_0, B_1, B_2, etc. which are known as Bernoulli's numbers, have the values

$$\begin{array}{llll} B_0 = 1, & B_1 = -\tfrac{1}{2}, & B_2 = 1/6, & B_3 = 0, \\ B_4 = -1/30, & B_5 = 0, & B_6 = 1/42, & B_7 = 0, \\ B_8 = 1/30, & B_9 = 0, & B_{10} = 5/66, & B_{11} = 0, \\ B_{12} = 691/2730 \text{ etc.} & & & \end{array}$$

That only even powers of z occur in the expansion (other than the term $-\frac{1}{2}z$) is clear from the fact that

$$\left(\frac{1}{e^z - 1} + \frac{1}{2}\right) z = \frac{(1 + e^z)z}{2(e^z - 1)} = \frac{(e^{-z} + 1)z}{2(1 - e^{-z})} = -\left(\frac{1}{e^{-z} - 1} + \frac{1}{2}\right) z$$

Since changing the sign of z in the foregoing expression does not change the sign of the function we see that the latter is even. Accordingly we may write

$$R = h(\mathrm{E}^r - 1)\left[\frac{1}{h\mathrm{D}} - \frac{1}{\mathrm{E} - 1} - \frac{1}{2}\right] f(a)$$

$$= h(\mathrm{E}^r - 1)\left[\frac{1}{h\mathrm{D}} - \frac{1}{e^{h\mathrm{D}} - 1} - \frac{1}{2}\right] f(a)$$

$$= h(\mathrm{E}^r - 1)\left[\frac{1}{h\mathrm{D}} - \frac{1}{2} - \left(\frac{1}{h\mathrm{D}} - \frac{1}{2} + \frac{B_2}{2\,!}h\mathrm{D} + \frac{B_4}{4\,!}h^3\mathrm{D}^3\right.\right.$$

$$\left.\left.\frac{B_6}{6\,!}h^5\mathrm{D}^5 \ldots\right)\right] f(a)$$

$$= -\frac{B_2}{2\,!}h^2[f'(a+rh) - f'(a)] - \frac{B_4}{4\,!}h^4[f'''(a+rh) - f'''(a)] - \ldots,$$

or
$$\frac{1}{h}\int_a^{a+rh} f(a)\,da = \tfrac{1}{2}[f(a) + f(a+rh)] + \sum_1^{r-1} f(a+rh)$$

$$-\frac{B_2}{2!}h[f'(a+rh) - f'(a)]$$

$$-\frac{B_4}{4!}h^3[f'''(a+rh) - f'''(a)]$$

$$-\frac{B_6}{6!}h^5[f^{(5)}(a+rh) - f^{(5)}(a)] + \ldots,$$

where, if $f(z)$ is a polynomial of finite degree, the derivatives on the right are carried forward until they become constant. The formula thus provides an expression for the integral of the function in terms of its values at intervals h throughout the range of the integration, and the successive derivatives at the beginning and end of the range. Alternatively, it provides an expression for the sum of the ordinates if the integral can itself be otherwise evaluated. It is not difficult to extend this formula to any function $f(z)$ for which the expression on the right does not terminate in general, but in which the remainder becomes negligible.

Verification

If $f(z) = z$, the expression gives

$$\frac{1}{2h}[(a+rh)^2 - a^2] = \tfrac{1}{2}(a + a + rh) + \sum_1^{r-1}(a+rh)$$

or
$$\frac{r^2h}{2} + ar = a + \frac{rh}{2} + (r-1)a + h\,\frac{r(r-1)}{2}$$

$$= \frac{r^2h}{2} + ar.$$

EXERCISES (V)

1. By taking $f(z) = z^2$ use the formula to determine B_1.
2. By taking $f(z) = z^3$ determine B_2.
3. Prove that $\sum_{1}^{\infty} e^{-r^2} = \dfrac{\sqrt{\pi} - 1}{2}$.
4. Prove that $\sum_{1}^{\infty} r^2 e^{-r^2} = \sqrt{\pi}/4$.
5. Prove that $\sum_{1}^{\infty} r^4 e^{-r^2} = \dfrac{1 . 3\sqrt{\pi}}{2^3}$.
6. Prove that $\sum_{1}^{\infty} r^{2n} e^{-r^2} = \dfrac{1 . 3 . 5 . . . (2n-1)\sqrt{\pi}}{2^{n+1}}$.
7. Prove that $\int_0^n f(x)\,dx = \sum_{r=0}^{n-1} f(r) + \frac{1}{2}[f(n) - f(0)] - \frac{1}{12}[f'(n) - f'(0)]$
$+ \frac{1}{720}[f'''(n) - f'''(0)] + \ldots$.

Throughout the whole of the foregoing discussion we have assumed that the function under consideration or the function, suitably transformed, can be expressed as a polynomial. For functions that are capable of expansion in a power series over a limited range this is a valid assumption *in practice*. It follows that since all problems of *interpolation* are concerned with the values of the function within such a limited range, the methods we have developed should suffice. For extrapolation, however, the problem is quite otherwise. It is obvious, for example, that if 2^x is tabulated from $x = 0$ to $x = 0{\cdot}1$ at intervals of 0·01 then the function within the range can, to a certain degree of accuracy, be represented by a polynomial and so interpolation be conducted with some assurance. If, however, this polynomial is used for extrapolation to, say, $x = 3$ it is extremely unlikely that the polynomial so found will lie close to the function 2^x as far as that value. Extrapolation in fact requires a scrutiny of the tabulated values from a rather different angle from that necessary for interpolation.

The Operator θ

Let us begin with an extended concept of differences. We write $\theta y(x) = y(x+1) - \lambda y(x)$ where λ is some number so far unspecified.

Thus
$$\theta y(x) = (E - \lambda)y(x),$$
$$\theta^2 y(x) = (E - \lambda)^2 y(x),$$
. . . .

and generally,

$$\theta^r y(x) = \left(E^r - r\lambda E^{r-1} + \frac{r(r-1)}{2!}\lambda^2 E^{r-2} + \ldots + (-1)^r \lambda^r\right) y(x)$$

$$= y(x+r) - r\lambda y(x+r-1) + \frac{r(r-1)}{2!}\lambda^2 y(x+r-2) + \ldots$$

$$+ (-1)^r \lambda^r y(x).$$

Accordingly we can draw up a table of extended differences where of course each column of differences will itself be a function of λ. Suppose it is possible to choose λ so as to reduce the rth column of extended differences, i.e. $\theta^r y(x)$, to zero or approximately zero. Then the functional form suggested must satisfy the equation

$$\theta^r y(x) = 0$$

or $(E - \lambda)^r y(x) = 0$,

i.e. $y(x) = \lambda^x[a_0 + a_1 x + \ldots + a_{r-1}x^{r-1}]$. (*See* Chapter 4.)

For purposes of extrapolation therefore this expression would replace the polynomial form that emerges out of the ordinary method of taking differences.

Example

x	$y(x)$	$\theta y(x)$	$\theta^2 y(x)$	$\theta^3 y(x)$
0	$-1{\cdot}0000$			
		λ		
1	$0{\cdot}0000$		$0{\cdot}75 - \lambda^2$	
		$0{\cdot}7500$		$1 - 2{\cdot}25\lambda + \lambda^3$
2	$0{\cdot}7500$		$1 - 1{\cdot}5\lambda$	
		$1 - 0{\cdot}75\lambda$		$0{\cdot}9375 - 3\lambda + 2{\cdot}25\lambda^2$
3	$1{\cdot}0000$		$0{\cdot}9375 - 2\lambda + 0{\cdot}75\lambda^2$	
		$0{\cdot}9375 - \lambda$		$0{\cdot}75 - 2{\cdot}8125\lambda + 3\lambda^2 - 0{\cdot}75\lambda^3$
4	$0{\cdot}9375$		$0{\cdot}75 - 1{\cdot}875\lambda + \lambda^2$	
		$0{\cdot}75 - 0{\cdot}9375\lambda$		
5	$0{\cdot}7500$			

The columns $\theta y(x)$ and $\theta^2 y(x)$ cannot be reduced to zero by any choice of λ. The column $\theta^3 y(x)$, on the other hand, reduces to zero for $\lambda = 0{\cdot}5$.

Accordingly, inserting this value of λ, the table becomes

x	y	$\theta y(x)$	$\theta^2 y(x)$	$\theta^3 y(x)$
0	$-1{\cdot}0000$			
		0·5		
1	0·0000		0·5	
		0·75		0
2	0·7500		0·250	
		0·625		0
3	1·0000		0·125	
		0·4375		0
4	0·9375		0·0625	
		0·2812		0
5	0·7500		0·0312	
		0·1718		
6	0·5469			

Now adding a zero to the θ^3 column below the line we have to add $\frac{1}{2}$[0·625] to the θ^2 column and so obtaining finally 0·5469 at $x = 6$. In this way we can extrapolate any distance without actually determining the form of the function.

In general the table takes the following form—

x	$y(x)$	$\theta y(x)$	$\theta^2 y(x)$
0	$y(0)$		
		$y(1) - \lambda y(0)$	
1	$y(1)$		$y(2) - 2\lambda y(1) + \lambda^2 y(0)$
		$y(2) - \lambda y(1)$	
2	$y(2)$		$y(3) - 2\lambda y(2) + \lambda^2 y(1)$
		$y(3) - \lambda y(2)$	
3	$y(3)$		.
		.	
.	.		.
		.	
$r-2$	$y(r-2)$		.
		$y(r-1) - \lambda y(r-2)$	
$r-1$	$y(r-1)$		$y(r) - 2\lambda y(r-1) + \lambda^2 y(r-2)$
		$y(r) - \lambda y(r-1)$	
r	$y(r)$		

The method is to inspect each column $\theta y(x)$, $\theta^2 y(x)$, etc., in turn to see if a value of λ can be found to make each entry in the column zero. Thus, the question we pose when inspecting the $\theta^2 y(x)$ column is, have the equations

$$y(2) - 2\lambda y(1) + \lambda^2 y(0) = 0,$$
$$y(3) - 2\lambda y(2) + \lambda^2 y(1) = 0,$$
$$\vdots$$
$$y(r) - 2\lambda y(r-1) + \lambda^2 y(r-2) = 0,$$

a solution in common?

This procedure is capable of immediate generalization

Let
$$\theta_1 y(x) = y(x+1) - \lambda_1 y(x) = (\mathrm{E} - \lambda_1)y(x),$$
$$\theta_2 y(x) = y(x+1) - \lambda_2 y(x) = (\mathrm{E} - \lambda_2)y(x),$$
$$\vdots$$
$$\theta_r y(x) = y(x+1) - \lambda_r y(x) = (\mathrm{E} - \lambda_r)y(x).$$

Then
$$\theta_1\theta_2\theta_3 \ldots \theta_r y(x) = \prod_1^r (\mathrm{E} - \lambda_r)y(x),$$

where any pair or more of the λ's may be equal.

Now suppose the table of extended differences is constructed to provide columns of values of $\theta_1 y(x)$, $\theta_2\theta_1 y(x)$, $\theta_3\theta_2\theta_1 y(x)$ etc. This means that the first column after that of $y(x)$ is constructed by taking $y(x+1) - \lambda_1 y(x)$ where λ_1 is unspecified. The second column is constructed from this one by taking λ_2 as a multiplier instead of λ_1 again, the next with λ_3, and so on. We now proceed to ask for what values of λ_1, λ_2, λ_3 etc. the extended differences in the rth column will be zero. Clearly this will occur for a function $y(x)$ which is such that

$$(\mathrm{E} - \lambda_r)(\mathrm{E} - \lambda_{r-1}) \ldots (\mathrm{E} - \lambda_2)(\mathrm{E} - \lambda_1)y(x) = 0.$$

If the λ's are all different this implies that $y(x)$ is the sum of r different exponentials. If two or more of the λ's are equal then corresponding to these terms a polynomial function will be multiplied into the corresponding exponential in the usual way (see Chapter 4). The procedure is best illustrated with an example.

Here six tabulated values are given and we proceed on the assumption therefore that y is expressible in the form

$$A_0\lambda_1^x + B_0\lambda_2^x + C_0\lambda_3^x.$$

x	y	$\theta_1 y$	$\theta_2\theta_1 y$	$\theta_3\theta_2\theta_1 y$
0	1·000			
		$3 - \lambda_1$		
1	3·000		$5 - 3(\lambda_1 + \lambda_2) + \lambda_1\lambda_2$	
		$5 - 3\lambda_1$		$8 - 5(\lambda_1 + \lambda_2 + \lambda_3) + 3(\lambda_1\lambda_2 + \lambda_2\lambda_3 + \lambda_3\lambda_1) - \lambda_1\lambda_2\lambda_3$
2	5·000		$8 - 5(\lambda_1 + \lambda_2) + 3\lambda_1\lambda_2$	
		$8 - 5\lambda_1$		$13{\cdot}5 - 8(\lambda_1 + \lambda_2 + \lambda_3) + 5(\lambda_1\lambda_2 + \lambda_2\lambda_3 + \lambda_3\lambda_1) - 3\lambda_1\lambda_2\lambda_3$
3	8·000		$13{\cdot}5 - 8(\lambda_1 + \lambda_2) + 5\lambda_1\lambda_2$	
		$13{\cdot}5 - 8\lambda_1$		$24{\cdot}25 - 13{\cdot}5(\lambda_1 + \lambda_2 + \lambda_3) + 8(\lambda_1\lambda_2 + \lambda_2\lambda_3 + \lambda_3\lambda_1) - 5\lambda_1\lambda_2\lambda_3$
4	13·500		$24{\cdot}25 - 13{\cdot}5(\lambda_1 + \lambda_2) + 8\lambda_1\lambda_2$	
		$24{\cdot}25 - 13{\cdot}5\lambda_1$		
5	24·250			

The $\theta_1 y$ column can obviously not be reduced to zero. If the $\theta_2\theta_1 y$ column is made zero for the first two terms the remaining terms are not zero. For the column $\theta_3\theta_2\theta_1 y$ solving for λ_1, $\lambda_1\lambda_2$, and $\lambda_1\lambda_2\lambda_3$ by equating the first three to zero, gives

$$\lambda_1 + \lambda_2 + \lambda_3 = 3{\cdot}5,$$

$$\lambda_1\lambda_2 + \lambda_2\lambda_3 + \lambda_3\lambda_1 = 3{\cdot}5,$$

$$\lambda_1\lambda_2\lambda_3 = 1.$$

Thus the λ's are the roots of the equation

$$\lambda^3 - 3{\cdot}5\lambda^2 + 3{\cdot}5\lambda - 1 = 0,$$

i.e. $\lambda_1 = 1, \quad \lambda_2 = 2, \quad \lambda_3 = 0{\cdot}5.$

We assume that the successive terms in the $\theta_3\theta_2\theta_1$ column are thereafter also zero.

Now reconstruct the extended difference table on this basis

x	y	$\lambda_1 = 1$ $\theta_1 y$	$\lambda_2 = 2$ $\theta_2\theta_1 y$	$\lambda_3 = \frac{1}{2}$ $\theta_3\theta_2\theta_1 y$
0	1			
		2		
1	3		−2	
		2		0
2	5		−1	
		3		0
3	8		−0·5	
		5·5		0
4	13·5		−0·25	
		10·75		0
5	24·25		−0·125	
		21·375		0
6	45·625		−0·0625	
		42·687		
7	88·312			

And clearly, by extending the column of zeros in $\theta_3\theta_2\theta_1 y$, we can extrapolate to any integral value of x. The cubic equation whose zeros are λ_1, λ_2, and λ_3 is

$$z^3 - (\lambda_1 + \lambda_2 + \lambda_3)z^2 + (\lambda_1\lambda_2 + \lambda_2\lambda_3 + \lambda_3\lambda_1)z - \lambda_1\lambda_2\lambda_3 = 0.$$

Associating this with the first three extended differences in the column $\theta_3\theta_2\theta_1 y$ equated to zero, and eliminating the terms in λ_1, λ_2, λ_3, we get

$$\begin{vmatrix} z^3 & z^2 & z & 1 \\ 8 & 5 & 3 & 1 \\ 13{\cdot}5 & 8 & 5 & 3 \\ 24{\cdot}25 & 13{\cdot}5 & 8 & 5 \end{vmatrix} = 0.$$

The solutions of this determine the three values of λ already found from which the extended difference table is constructed.

These calculations have been performed without attempting to find a formal expression for $y(x)$. Actually this formal expression would be suitable in practice for interpolating or extrapolating to a point intermediate between two integral values of x. It is found directly as follows—

$$y(a + xh) = \mathrm{E}^x y(a).$$

Write $\mathrm{E}^x = a_0 + a_1(\mathrm{E} - \lambda_1) + a_2(\mathrm{E} - \lambda_1)(\mathrm{E} - \lambda_2) + \ldots,$

where a_0, a_1, a_2 etc. are functions of x to be determined. This series terminates if $y(x)$ is in fact expressible as a finite set of exponentials.

Inserting $\mathrm{E} = \lambda_1$ leads to $a_0 = \lambda_1^x$,

$$\mathrm{E} = \lambda_2 \text{ gives } a_1 = \frac{\lambda_1^x}{\lambda_1 - \lambda_2} + \frac{\lambda_2^x}{\lambda_2 - \lambda_1},$$

leading to

$$a_2 = \frac{\lambda_1^x}{(\lambda_1 - \lambda_2)(\lambda_1 - \lambda_3)} + \frac{\lambda_2^x}{(\lambda_2 - \lambda_1)(\lambda_2 - \lambda_3)} + \frac{\lambda_3^x}{(\lambda_3 - \lambda_1)(\lambda_3 - \lambda_2)}$$

$$a_3 = \frac{\lambda_1^x}{(\lambda_1 - \lambda_2)(\lambda_1 - \lambda_3)(\lambda_1 - \lambda_4)} + \frac{\lambda_2^x}{(\lambda_2 - \lambda_1)(\lambda_2 - \lambda_3)(\lambda_2 - \lambda_4)}$$
$$+ \frac{\lambda_3^x}{(\lambda_3 - \lambda_1)(\lambda_3 - \lambda_2)(\lambda_3 - \lambda_4)} + \frac{\lambda_4^x}{(\lambda_4 - \lambda_1)(\lambda_4 - \lambda_2)(\lambda_4 - \lambda_3)}$$

and so on.

Thus

$$y(a + xh) = \lambda_1^x \left[1 + \frac{\theta_1}{\lambda_1 - \lambda_2} + \frac{\theta_1\theta_2}{(\lambda_1 - \lambda_2)(\lambda_1 - \lambda_3)} + \ldots\right] y(a)$$
$$+ \lambda_2^x \left[\frac{\theta_1}{\lambda_2 - \lambda_1} + \frac{\theta_1\theta_2}{(\lambda_2 - \lambda_1)(\lambda_2 - \lambda_3)} + \ldots\right] y(a)$$
$$+ \lambda_3^x \left[\frac{\theta_1\theta_2}{(\lambda_3 - \lambda_1)(\lambda_3 - \lambda_2)}\right.$$
$$\left. + \frac{\theta_1\theta_2\theta_3}{(\lambda_3 - \lambda_1)(\lambda_3 - \lambda_2)(\lambda_3 - \lambda_4)} + \ldots\right] y(a) + \ldots$$

which may be regarded as an extension of the Gregory–Newton formula in ordinary differences.

Example

x	y	$\theta_1 y$	$\theta_2\theta_1 y$
0	9		
		$15 - 9\lambda_1$	
1	15		$21 - 15(\lambda_1 + \lambda_2) + 9\lambda_1\lambda_2$
		$21 - 15\lambda_1$	
2	21		$15 - 21(\lambda_1 + \lambda_2) + 15\lambda_1\lambda_2$
		$15 - 21\lambda_1$	
3	15		

We easily find that $\lambda_1 = 2$, $\lambda_2 = 3$, make the last column zero.
Hence

x	y	$\theta_1 y$	$\theta_2\theta_1 y$
0	9		
		-3	
1	15		0
		-9	
2	21		0
		-27	
3	15		0
		-81	
4	-51		

Here $\quad y(0) = 9, \quad \theta_1 y(0) = -3, \quad \theta_2\theta_1 y(0) = 0.$

Inserting these in the general expression for y gives therefore

$$y(x) = 2^x\left[9 - \frac{3}{2-3}\right] + 3^x\left[\frac{-3}{3-2}\right] = 12 \,.\, 2^x - 3^{x+1}.$$

Verification

$$x = 0 \quad y(0) = 12 - 3 = 9.$$
$$x = 1 \quad y(1) = 24 - 9 = 15.$$
$$x = 2 \quad y(2) = 48 - 27 = 21.$$

Hence we can use this formula for interpolation, or extrapolation, at intermediate values of x.

EXERCISES (VI)

1. Show that when $\lambda_1 = \lambda_2 = \lambda_3 = \ldots \lambda_r = \lambda$

$$y(a + xh) = \lambda^x \left[1 + \frac{x\theta}{\lambda} + \frac{x(x-1)\theta^2}{2!\lambda^2} + \ldots + \frac{x(x-1)\ldots(x-r+1)\theta^r}{r!\lambda^r}\right] y(a).$$

2. Apply the method of extended differences to find the value of y at $x = 8$ in the following case—

x	0	1	2	3	4	5	6
y	8	19	42	83	128	45	-818

3. Find a formal expression for $y(x)$ for the following table—

x	0	1	2	3	4	5
y	1	4	12	34	96	274

In general if the tabulated values are $2n$ in number

$$\begin{array}{cccccc} x & 0 & 1 & 2 & \ldots & 2n-1 \\ y & y_0 & y_1 & y_2 & \ldots & y_{2n-1}, \end{array}$$

and the expression for y is of the form

$$A_0\lambda_1^x + B_0\lambda_2^x + \ldots + N_0\lambda_n^x,$$

the equation whose solutions are $\lambda_1, \lambda_2, \ldots, \lambda_n$, is

$$\begin{vmatrix} 1 & z & z^2 & \ldots & z^n \\ y_0 & y_1 & y_2 & \ldots & y_n \\ y_1 & y_2 & y_3 & \ldots & y_{n+1} \\ \cdot & \cdot & \cdot & \ldots & \cdot \\ \cdot & \cdot & \cdot & \ldots & \cdot \\ \cdot & \cdot & \cdot & \ldots & \cdot \\ y_{n-1} & y_n & y_{n+1} & \ldots & y_{2n-1} \end{vmatrix} = 0.$$

If the values of λ so found are not distinct then in place of separate exponential terms there occur polynomial terms in association with the exponential terms corresponding to these multiple values of λ.

If there is an odd number of tabulated values, say $2n + 1$, then clearly it is not possible to fit either a set of n or $(n + 1)$ exponential terms. There are too many values for the one case and too few for the other. To make the number of unknown constants equal to the given number of values, viz. $2n + 1$, at least two of the values of λ

must be equal. This means that there must be at least one term present of the form $\lambda^x P(x)$ where $P(x)$ is a polynomial. Since the order of occurrence of the λ's in the function is arbitrary we begin by making $\lambda_1 = \lambda_2$ and proceed as before. There are then sufficient terms to determine all the λ's, although of course it may turn out that $\lambda_r = \lambda_s = \lambda_t = \lambda_1$ say. The procedure, nevertheless, will not be affected by this fact.

There are many alternative methods of handling the case of an odd number of values. We may, for example, assume that y takes the form $\Sigma A_r \lambda_r^x + K$ where K is itself to be determined. In effect this is equivalent to writing one of the exponential terms as $K(1)^x$. Each entry is then reduced by K and the procedure follows as before to find $\lambda_1, \lambda_2 \ldots$ and K.

EXERCISE (VII)

Show that if $y_0, y_1, y_2, \ldots$ are a series of values corresponding to $x = 0, 1, 2, \ldots,$ which are exactly represented by r exponential terms in which r is unknown, then r can be found by evaluating in succession the determinants

$$\begin{vmatrix} y_0 & y_1 \\ y_1 & y_2 \end{vmatrix}, \quad \begin{vmatrix} y_0 & y_1 & y_2 \\ y_1 & y_2 & y_3 \\ y_2 & y_3 & y_4 \end{vmatrix}, \ldots .$$

If the determinant of the mth order is zero then $r = m - 1$.

Differentiation and Integration by Extended Differences

It remains to determine formulae for the differential coefficient and the definite integral of a tabulated function in terms of extended differences. From the definition we have

$$\theta_1 = \mathrm{E} - \lambda_1,\ \theta_2 = \mathrm{E} - \lambda_2,\ \ldots,\ \theta_r = \mathrm{E} - \lambda_r.$$

For the moment we shall assume that the constants $\lambda_1, \lambda_2, \ldots, \lambda_r$ are distinct. Since $\mathrm{E} = e^{h\mathrm{D}}$ and $h\mathrm{D} = \log \mathrm{E}$ we write operationally

$$\begin{aligned} h^n\mathrm{D}^n = (\log \mathrm{E})^n &= a_0 + a_1\theta_1 + a_2\theta_1\theta_2 + a_3\theta_1\theta_2\theta_3 + \ldots \\ &= a_0 + a_1(\mathrm{E} - \lambda_1) + a_2(\mathrm{E} - \lambda_1)(\mathrm{E} - \lambda_2) \\ &\quad + a_3(\mathrm{E} - \lambda_1)(\mathrm{E} - \lambda_2)(\mathrm{E} - \lambda_3) + \ldots \end{aligned}$$

where n is a positive or a negative integer. Integration corresponds of course to $n = -1$. If the original tabulated function is capable of representation in terms of r exponentials, i.e. if the rth extended differences are zero so that $\theta_1\theta_2 \ldots \theta_r y(x) = 0$, i.e.

$$(\mathrm{E} - \lambda_1)(\mathrm{E} - \lambda_2) \ldots (\mathrm{E} - \lambda_r)y(x) = 0,$$

then the series on the right terminates with the term

$$a_{r-1}(\mathrm{E} - \lambda_1) \ldots (\mathrm{E} - \lambda_{r-1}).$$

The coefficients $a_0, a_1, \ldots, a_{r-1}$ are now easily determined in succession.

Replacing E by λ_1, λ_2, and λ_3 respectively, gives at once

$$a_0 = (\log \lambda_1)^n,$$

$$a_1 = [(\log \lambda_2)^n - (\log \lambda_1)^n]/(\lambda_2 - \lambda_1),$$

$$a_2 = -\frac{(\lambda_1 - \lambda_2)(\log \lambda_3)^n + (\lambda_2 - \lambda_3)(\log \lambda_1)^n + (\lambda_3 - \lambda_1)(\log \lambda_2)^n}{(\lambda_1 - \lambda_2)(\lambda_2 - \lambda_3)(\lambda_3 - \lambda_1)}$$

etc.

Thus, for example, if the third extended differences are zero then the above series finishes at the third term so that

$$hy'(x) = (\log \lambda_1)y(x) + \frac{\log \lambda_2 - \log \lambda_1}{\lambda_2 - \lambda_1}\theta_1 y(x)$$
$$- \frac{(\lambda_1 - \lambda_2)\log \lambda_3 + (\lambda_2 - \lambda_3)\log \lambda_1 + (\lambda_3 - \lambda_1)\log \lambda_2}{(\lambda_1 - \lambda_2)(\lambda_2 - \lambda_3)(\lambda_3 - \lambda_1)}\theta_1\theta_2 y(x).$$

Moreover, since

$$\int_a^{a+rh} y(x)\,dx = \frac{(\mathrm{E}^r - 1)}{h\mathrm{D}}y(a),$$

it follows that

$$\int_a^{a+rh} y(x)\,dx = \frac{1}{\log \lambda_1}[y(a + rh) - y(a)]$$
$$+ \frac{\log \lambda_1 - \log \lambda_2}{(\lambda_2 - \lambda_1)\log \lambda_1 \log \lambda_2}\theta_1[y(a + rh) - y(a)]$$
$$- \frac{(\lambda_1 - \lambda_2)\log \lambda_1 \log \lambda_2}{(\lambda_1 - \lambda_2)(\lambda_2 - \lambda_3)(\lambda_3 - \lambda_1)\log \lambda_1 \log \lambda_2 \log \lambda_3}$$
$$\times\, \theta_1\theta_2[y(a + rh) - y(a)] + \ldots.$$

Example

x	y	$\theta_1 y$	$\theta_1\theta_2 y$	$y'(x)$
1	9			$12 \log 2 - 3 \log 3$
		-3		
2	15		0	$24 \log 2 - 9 \log 3$
		-9		
3	21		0	$48 \log 2 - 27 \log 3$
		-27		
4	15		0	$96 \log 2 - 54 \log 3$
		-81		
5	-51		0	
		-243		
6	-345		0	
		-729		
7	-1419			
		-2187		
8	-5025			

From the table for y it is easily found that $\lambda_1 = 2$ and $\lambda_2 = 3$. The table of extended differences has been constructed on this basis. The values are given for y as far as $x = 5$. Thereafter it has been extended from the table as far as $x = 8$. (An ordinary polynomial interpolation would have given $y = -1629$ at $x = 8$.)

Since $h = 1$ we now have

$$y'(1) = (\log 2)y(1) + \frac{\log 3 - \log 2}{3 - 2}\,\theta_1 y(1)$$

$$= 9 \log 2 - 3 \log 3/2 = 12 \log 2 - 3 \log 3.$$

The remaining values of $y'(x)$ have been calculated in the same way.

Again to evaluate $I = \int_1^5 y(x)\,dx$

$$I = \frac{1}{\log 2}[-51 - 9] + \frac{\log 2 - \log 3}{\log 2 \log 3}[-243 + 3] = \frac{180}{\log 2} - \frac{240}{\log 3}.$$

The various cases we have dealt with have really been systematic, but special procedures for transforming a given table of entry, in which no trend is immediately discernible, to one in which the trend is obvious must be devised. The process has been tentative, in the sense that we have examined whether the data can be shown numerically to satisfy equations of the type

$$\Delta^r y(x) = 0,$$

in which case $y(x)$ is expressible as a polynomial function of x, or of the type

$$(E - \lambda_1)(E - \lambda_2) \ldots (E - \lambda_r)y(x) = 0,$$

in which case $y(x)$ is expressible as a series of exponentials. In this case modifications are obvious when two or more of the λ's are equal.

More generally we have had in mind finding relations of the form

$$y(x + r) = \varphi[y(x), y(x + 1), \ldots, y(x + r - 1)]$$

which once established as covering the data can plausibly be assumed for purposes of extrapolation. For interpolation purposes, however, it would be necessary, if this trend were to be used, to find $y(x)$ explicitly from the above as a function of x. The above relation is a difference equation and the search for $y(x)$ is the search for its solution. To such questions we shall turn shortly. There are therefore many other approaches to the problem of transforming a set of data to one in which the trend is obvious than those exemplified in the illustrations above.

CHAPTER 3

The Determination of Difference Equations

In this chapter we do not propose yet to make any systematic study of how to obtain solutions of difference equations. We intend rather to show how such equations arise.

What is a Difference Equation?

A difference equation is a relation between the differences of a function at one or more general values of the independent variable. Thus the following are difference equations—

$$\Delta y(x+1) - \Delta^2 y(x-1) - 1 = 0,$$
$$[\Delta^2 y(x)]^2 - y(x+2) + 2 = 0.$$

Since the rth difference of the functions $y(x)$, i.e. $\Delta^r y(x)$, can always be expressed in terms of $y(x)$, $y(x+1)$, . . ., $y(x+r)$, it is possible to set out a difference equation in an alternative way not directly involving the symbol Δ from which the term *difference* equation has arisen. Thus, in the first of the illustrations above, since

$$\Delta y(x+1) = y(x+2) - y(x+1)$$

and

$$\Delta^2 y(x-1) = y(x+1) - 2y(x) + y(x-1)$$

the equation may take the alternative form

$$y(x+2) - y(x+1) = y(x-1) - 2y(x) + y(x-1) + 1$$

or

$$0 = y(x+2) - 2y(x+1) + 2y(x) - y(x-1).$$

Generally therefore, any functional relation of the form

$$F[x, y(x), y(x+1), \ldots, y(x+r)] = 0 \tag{1}$$

is a difference equation. Such expressions are frequently found also under the name of recurrence relations. When an expression for

$y(x)$ has been found to satisfy (1) it is called a solution of the difference equation.

How Difference Equations Arise

We begin by examining a number of ways in which difference equations arise.

Difference Equations Arising from Generating Functions

If $f(z)$ can be expressed as a power series absolutely convergent within a given range,

$$f(z) = y_0 + y_1 z + y_2 z^2 + \ldots + y_n z^n + \ldots$$

or if $f(z) = a_0 + a_1 \cos z + a_2 \cos 2z + \ldots + a_n \cos nz + \ldots$
$$+ b_1 \sin z + b_2 \sin 2z + \ldots + b_n \sin nz + \ldots,$$

i.e. as a Fourier expansion, then y_n, a_n, b_n are functions of n and the function $f(z)$ is said to be the generating function of y_n, a_n, and b_n.

It frequently happens, as we shall see, that the determination of such functional coefficients depends on solving a difference equation.

Examples

1. Suppose $e^z/(1 - z)$ is expanded in ascending powers of z, viz.

$$e^z/(1 - z) = y_0 + y_1 z + y_2 z^2 + \ldots + y_n z^n + \ldots$$

which we assume to be absolutely convergent for $|z| < 1$. Multiply both sides by $(1 - z)$. The coefficient of z^n on the right-hand side is then

$$y_n - y_{n-1}.$$

Since the coefficient of z^n in the expansion of e^z is $1/n!$, on equating these equivalent coefficients we have

$$y_n - y_{n-1} = 1/n!$$

Thus
$$\Delta y_n = 1/(n + 1)!$$

Remembering the meaning attached to Δ^{-1}, this gives

$$y_n = A + \sum_{1}^{n-1} 1/(n + 1)!$$

where A is an unknown constant. In addition, it is clear by actually expanding to the first few terms by Taylor's theorem that $y_0 = 1$, $y_1 = 2$, $y_2 = 5/2$ etc. Inserting this value for y_2 in the expression for y_n, we obtain

$$5/2 = A + \sum_{1}^{1} 1/(n+1)! = A + \tfrac{1}{2}.$$

Thus $$A = 2.$$

Accordingly,

$$\begin{aligned} y_n &= 2 + \sum_{1}^{n-1} 1/(n+1)! \\ &= 1/0! + 1/1! + 1/2! + \ldots + 1/n!, \end{aligned}$$

the first $(n + 1)$ terms in the expansion for e. Note that this is consistent with $y_0 = 1$, $y_2 = 2$, etc. As n tends to infinity, y_n tends to e. This justifies the assumption of absolute convergence, after the fact.

We notice that in this case the determination of y_n has involved solving the difference equation

$$y_n - y_{n-1} = 1/n!$$

in association with a particular value, viz. $y_2 = 5/2$.

2. Consider the expansion,

$$(6 - z^2)/(z^2 - 3z + 2) = y_0 + y_1 z + y_2 z^2 + \ldots + y_n z^n + \ldots,$$

where, as before, the series is assumed to be absolutely convergent for $|z| < 1$. Multiplying both sides by $(z^2 - 3z + 2)$ and equating the coefficients of z^n on both sides, we easily obtain the relation

$$0 = 2y_{n+2} - 3y_{n+1} + y_n,$$

where $y_0 = 3$, $y_1 = 9/2$, $y_2 = 19/4 \ldots$. Again we have a difference equation. We shall shortly examine equations of this type in general; here we shall solve this particular equation by a special method. We write,

$$\begin{aligned} 2y_{n+2} - 3y_{n+1} + y_n &= 2(y_{n+2} - y_{n+1}) - (y_{n+1} - y_n) \\ &= 2\Delta y_{n+1} - \Delta y_n \\ &= \Delta(2y_{n+1} - y_n). \end{aligned}$$

It follows that $$0 = \Delta(2y_{n+1} - y_n),$$

i.e. $$2y_{n+1} - y_n = A, \text{ a constant.} \tag{2}$$

Thus, $$2y_1 - y_0 = A$$
$$A = 9 - 3 = 6.$$

We are now faced with the problem of finding y_n from (2). To this end let us try to reduce the left-hand side to a complete difference. It is not difficult to see that this can be done by multiplying throughout by 2^n. Thus,

$$2^{n+1}y_{n+1} - 2^n y_n = 6 \,.\, 2^n,$$

or $$\Delta(2^n y_n) = 6 \,.\, 2^n.$$

Accordingly, $$2^n y_n = \Delta^{-1}(6 \,.\, 2^n) = B + 6 \sum_1^{n-1} 2^n,$$

and $$y_n = B \,.\, 2^{-n} + 6 \,.\, 2^{-n} \sum_1^{n-1} 2^n.$$

To determine the constant B, we put $n = 2$,

$$y_2 = 19/4 = B/4 + 6\sum_1^1 2^n/4$$

or $$19 = B + 12,$$

thus $$B = 7,$$

and $$y_n = 7 \,.\, 2^{-n} + 6 \,.\, 2^{-n} \sum_1^{n-1} 2^n$$

or $$y_n = 6 - 5 \,.\, 2^{-n}.$$

Again the assumption of absolute convergence is justified *a posteriori*.

3. Consider the expansion

$$\frac{2 + \cos z}{5 + 4 \cos z} = a_0 + a_1 \cos z + a_2 \cos 2z + \ldots + a_n \cos nz + \ldots.$$

A cosine series is chosen as the left-hand side is an even function of z. Again assume this series to be absolutely convergent.

Multiplying both sides of the above relation by $(5 + 4 \cos z)$ and remembering that

$$2 \cos^2 z = 1 + \cos 2z,$$
$$2 \cos z \,.\, \cos nz = \cos (n - 1)z + \cos (n + 1)z,$$

on collecting like terms, we find,

$$2 + \cos z = 5a_0 + 2a_1 + (4a_0 + 5a_1 + 2a_2) \cos z +$$
$$+ \ldots + (5a_n + 2a_{n-1} + 2a_{n+1}) \cos nz + \ldots.$$

Equating coefficients of cos nz,

$$2 = 5a_0 + 2a_1 \tag{3}$$

$$1 = 4a_0 + 5a_1 + 2a_2 \tag{4}$$

$$0 = 5a_n + 2a_{n-1} + 2a_{n+1} \tag{5}$$

Equation (5) is in fact the difference equation from which a general expression for a_n may be obtained, which in this case is required to be consistent with (3) and (4) and must satisfy the condition for the convergence of the original series.

It can easily be verified that

$$a_n = A(-2)^n + B(-\tfrac{1}{2})^n \tag{6}$$

satisfies (5) where A and B are arbitrary constants. In order that the series may be convergent, the constant A must be identically zero. Thus

$$a_0 = B, \qquad a_1 = -B/2, \qquad \text{and} \qquad a_2 = B/4.$$

It merely remains to show that (3) and (4) are consistent with $A = 0$; (3) becomes

$$2 = 5B - B,$$

i.e.

$$B = \tfrac{1}{2},$$

and (4) becomes

$$1 = 4B - 5B/2 + B/2,$$

i.e.

$$B = \tfrac{1}{2}.$$

Thus all the relations are consistent and we have the expansion

$$\frac{2 + \cos z}{5 + 4 \cot z} = \tfrac{1}{2}[1 - \tfrac{1}{2}\cos z + (\tfrac{1}{2})^2 \cos 2z + \ldots + (-1)^n(\tfrac{1}{2})^n \cos nz + \ldots].$$

EXERCISE (I)

By replacing cos nz by the real part of e^{inz} sum the series on the right and so verify the expression on the left.

4. Let $te^{xt}/(e^t - 1)$ be expanded in ascending powers of t, and let the coefficient of $t^r/r!$ in the expansion be $B_r(x)$ so that

$$\frac{te^{xt}}{e^t - 1} = \sum_{r=0}^{\infty} \frac{t^r}{r!} B_r(x).$$

Taking the differences of both sides with respect to x

$$\frac{t}{e^t - 1}\left[e^{(x+1)t} - e^{xt}\right] = te^{xt} = \sum_0^\infty \frac{t^r}{r!}\Delta B_r(x)$$

$$= t + \frac{xt^2}{1!} + \frac{x^2t^3}{2!} + \ldots + \frac{x^{r-1}t^r}{(r-1)!} + \ldots,$$

and equating powers of t, we have

$$B_r(x+1) - B_r(x) = rx^{r-1}.$$

It is not difficult to see that a polynomial of degree r in x may be found to satisfy this difference equation, for if we write

$$B_r(x) = a_0 + a_1x + a_2x^2 + \ldots + a_rx^r,$$

there being $r + 1$ coefficients, then, on inserting this expression into the difference equation and equating coefficients we obtain r linear equations to determine $a_1, a_2, \ldots, a_r$ uniquely. The coefficient a_0 is, however, arbitrary so far, but since $a_0 = B_r(0)$ it has a definite significance, for $B_r(0)$, or simply B_r, is the rth *Bernoulli number*, by definition (*see* Chapter 2). The Bernoulli numbers are the coefficients of $t^r/r!$ in the expansion of $t/(e^t - 1)$ in ascending powers of t, on inserting $x = 0$ in the definition of $B_r(x)$. If we write

$$\frac{t}{e^t - 1} = \frac{t}{t + t^2/2! + t^3/3! + \ldots} = \left(1 + \frac{t}{2!} + \frac{t^2}{3!} + \ldots\right)^{-1},$$

and expand in ascending powers of t, it is easily seen that—

$$B_0 = 1, \qquad B_1 = -\tfrac{1}{2}, \qquad B_2 = 1/6, \qquad B_3 = 0,$$

$$B_4 = -1/30, \qquad B_5 = 0, \qquad B_6 = 1/42, \text{ etc.},$$

the numbers with the odd suffixes always vanishing after B_1.

The polynomials $B_r(x)$ are known as *Bernoulli polynomials* and are uniquely specified, for, with these conditions, we find that

$$B_r(x+1) - B_r(x) = rx^{r-1}$$

is satisfied by

$$B_0(x) = 1,$$

$$B_1(x) = x - \tfrac{1}{2},$$

$$B_2(x) = x^2 - x + 1/6,$$

$$B_3(x) = x^3 - 3x^2/2 + x/2,$$

$$B_4(x) = x^4 - 2x^3 + x^2 - 1/30,$$
$$B_5(x) = x^5 - 5x^4/2 + 5x^3/3 - x/6,$$
$$B_6(x) = x^6 - 3x^5 + 5x^4/2 - x^2/2 + 1/42, \text{ etc.}$$

The Bernoulli numbers, as we have already seen, enter into such expressions as the Euler-MacLaurin summation and integration formulae. In association with these numbers therefore, for each n, the Bernoulli polynomials provide solutions of the difference equation—

$$y(x+1) - y(x) = nx^{r-1}.$$

It follows immediately that if $P(x)$ is any polynomial then the solution of the difference equation

$$y(x+1) - y(x) = P(x)$$

can be expressed in terms of Bernoulli polynomials.

Difference Equations Arising from Differential Equations

This process is best illustrated through an example. Consider the second order differential equation

$$(6x^2 - 5x + 1)y'' + 2(12x - 5)y' + 12y = 0,$$

given that $y(0) = 1$ and $y'(0) = 0$.

To construct the series for $y(x)$ we require the values of y', y'', y''' etc. at the point $x = 0$. Write

$$y(x) = y_0 + y_1x + y_2x^2 + \ldots + y_nx^n + \ldots,$$

where y_n is in fact the value of the nth differential coefficient of y divided by $n!$ when $x = 0$, i.e. $y_n = y^n(0)/n!$. Then

$$y'(x) = y_1 + 2y_2x + 3y_3x^2 + \ldots + ny_nx^{n-1} + \ldots,$$

and $$y''(x) = 2y_2 + 3 \cdot 2y_3x + \ldots + n(n-1)y_nx^{n-2} + \ldots.$$

On multiplying the series for y by 12, the series for y' by $2(12x - 5)$, and the series for y'' by $(6x^2 - 5x + 1)$, adding these resultant series together and equating the coefficients of x^n to zero, we obtain

$$(n+2)(n+1)y_{n+2} - 5(n+1)ny_{n+1} + 6n(n-1)y_n - 10(n+1)y_{n+1} + 24ny_n + 12y_n = 0,$$

or

$$(n+2)(n+1)y_{n+2} - 5(n+2)(n+1)y_{n+1} + 6(n+2)(n+1)y_n = 0,$$

i.e. $$y_{n+2} - 5y_{n+1} + 6y_n = 0 \tag{7}$$

which is, of course, a difference equation. With this we must associate

$$y(0) = 1 \text{ and } y'(0) = 0, \text{ i.e. } y_0 = 1 \text{ and } y_0' = 0.$$

So we see that the solution in series of this particular differential equation rests on the solution of the above difference equation. For the moment, we are concerned merely to recognize how difference equations arise. In the next chapter we shall proceed to a systematic study of how to solve such equations. In order to complete this example, however, we ask the reader to accept that the solution of equation (7) is of the form

$$y_n = A \,.\, 3^n + B \,.\, 2^n$$

where A and B are constants. This can be verified by direct insertion.

Now, $y_0 = 1$,

thus $1 = A + B$,

and $y_1 = 0$,

thus $0 = 3A + 2B$.

Hence $A = -1/2$

and $B = 3/2$.

Thus $y_n = (-1/2) \,.\, 3^n + (3/2) \,.\, 2^n$,

hence $y(x) = \sum_{n=0}^{\infty} [-3^n/2 + 3 \,.\, 2^{n-1}]x^n$.

The convergence of this series for the range $|x| < 1/3$ can easily be verified.

Difference Equations Arising in the Direct Statement of a Problem

The classical method of proving that there is an unlimited number of primes is as follows—

Suppose the contrary be true and that the limited system of primes is

$$y_1, y_2, y_3, \ldots, y_n,$$

where $y_1 < y_2 < y_3 \ldots < y_n$.

Then the number N where

$$N = y_1 y_2 y_3 \dots y_n + 1$$

is prime to $y_r (r = 1, 2, 3, \dots, n)$ since on dividing N by y_r, the remainder is 1. Hence N is a prime which is greater than y_n; accordingly if y_n were the greatest prime there would be a greater one. From this contradiction it follows that there is no greatest prime number.

Let us write this process of derivation of "greater primes" from "lesser primes" thus—

$$y_n = 1 + y_1 y_2 y_3 \dots y_{n-1}.$$

Then $$y_{n+1} = 1 + y_1 y_2 y_3 \dots y_n.$$

Hence, $$\frac{y_{n+1} - 1}{y_n - 1} = \frac{y_1 y_2 y_3 \dots y_n}{y_1 y_2 y_3 \dots y_{n-1}} = y_n$$

or $$y_{n+1} - 1 = y_n^2 - y_n = (y_n - \tfrac{1}{2})^2 - \tfrac{1}{4}.$$

Thus the problem gives rise to the above, a difference equation in y_n, which by writing $z_n = y_n - \frac{1}{2}$ takes the more compact form

$$z_{n+1} = z_n^2 + \tfrac{1}{4}.$$

As we shall see, this is a non-linear difference equation of the first order and of the second degree.

We do not propose here to attempt to find the expression for z_n. The equation is not one of the type that can be solved by any of the simpler methods. Further illustrations of how difference equations arise will be dealt with in Chapter 7 after various methods of solution have been more systematically studied.

EXERCISES (II)

1. Assuming that

$$\frac{\log(1 + z)}{(1 - z)} = y_0 + y_1 z + y_2 z^2 + \dots + y_n z^n + \dots,$$

show that y_n satisfies the difference equation

$$\Delta y_n = (-1)^n/(n + 1),$$

with $$y_2 = 1/2.$$

Verify that $$y_n = \sum_{1}^{n-1} (-1)^n/(n + 1) + 1.$$

2. If $\dfrac{\sin z}{2 + z^2} = \sum_{n=0}^{\infty} y_n z^n,$

show that
$$2y_{2n} + y_{2n-2} = 0,$$
and
$$2y_{2n+1} + y_{2n-1} = (-1)^n/(2n+1)!$$
where
$$n = 1, 2, 3, \ldots.$$

3. Show that $xy'' + y' + 2y = 0$,

and
$$y(x) = y_0 + y_1x + \ldots + y_nx^n + \ldots,$$
where
$$y(0) = -\tfrac{1}{2},$$
lead to
$$(n+1)^2y_{n+1} + 2y_n = 0.$$
Verify that
$$y(x) = -\tfrac{1}{2} + \sum_{n=0}^{\infty}[(-2)^n/(n!)^2]x^n$$
satisfies the differential equation.

Formal Derivation of Difference Equations

Let $y_n = f(n, A)$, where A is a constant. Then

$$y_{n+1} = f(n+1, A).$$

On eliminating A between these two equations we obtain a relation between n, y_n, and y_{n+1}, say

$$F(n, y_n, y_{n+1}) = 0.$$

This is a difference equation of the first order, the order referring to the interval between the largest suffix attached to y, viz. $(n+1)$ and the smallest suffix, viz. n.

More generally, when

$$y_n = f(n, A_1, A_2, \ldots, A_r),$$
then
$$y_{n+1} = f(n+1, A_1, A_2, \ldots, A_r),$$
$$\cdots\cdots\cdots\cdots\cdots\cdots$$
$$y_{n+r} = f(n+r, A_1, \ldots, A_r).$$

These are $(r+1)$ equations from which we may suppose A_1, $A_2, \ldots A_r$ can be eliminated leading to a relation of the form

$$F(n, y_n, y_{n+1}, \ldots, y_{n+r}) = 0.$$

This is a difference equation of the rth order, since the highest suffix attached to y is $(n+r)$ and the lowest is n.

Examples

1. Let $y_n = An + A^2$, where A is a constant. Then

$$y_{n+1} = A(n+1) + A^2,$$

Hence $$A = y_{n+1} - y_n.$$

Accordingly, eliminating A, the difference equation is

$$y_n = n(y_{n+1} - y_n) + (y_{n+1} - y_n)^2$$

or $$y^2_{n+1} + y_n^2 - 2y_n y_{n+1} + ny_{n+1} - (n+1)y_n = 0.$$

2. $y_n = A^{2^n}$ where A is a constant.

$$y_{n+1} = A^{2 \cdot 2^n} = y_n^2.$$

This means that the solution of the difference equation $y_{n+1} = y_n^2$ is $y_n = A^{2^n}$ where A is a constant.

3. Let $y_n = A_1 . 2^n + A_2 . 3^n$ where A_1 and A_2 are arbitrary constants. Then

$$y_{n+1} = 2A_1 . 2^n + 3A_2 . 3^n,$$
$$y_{n+2} = 4A_1 . 2^n + 9A_2 . 3^n.$$

Hence, the eliminant of A_1 and A_2 being

$$\begin{vmatrix} y_n & 1 & 1 \\ y_{n+1} & 2 & 3 \\ y_{n+2} & 4 & 9 \end{vmatrix} = 0,$$

on expanding the determinant, we obtain

$$y_{n+2} - 5y_{n+1} + 6y_n = 0,$$

a difference equation of the second order.

Alternative Method

$$y_n = A_1 2^n + A_2 3^n$$
$$(\mathrm{E} - 2)y_n = A_1(2^{n+1} - 2 . 2^n) + A_2(3^{n+1} - 2 . 3^n)$$
$$= A_2 3^n$$
$$(\mathrm{E} - 3)(\mathrm{E} - 2)y_n = A_2(3^{n+1} - 3 . 3^n) = 0$$
$$(\mathrm{E}^2 - 5\mathrm{E} + 6)y_n = 0,$$

or the required difference equation is

$$y_{n+2} - 5y_{n+1} + 6y_n = 0.$$

4.
$$y_n = A_1 + (-1)^n A_2 + A_3 n$$
$$y_{n+1} = A_1 - (-1)^n A_2 + A_3 n + A_3$$
$$y_{n+2} = A_1 + (-1)^n A_2 + A_3 n + 2A_3$$
$$y_{n+3} = A_1 - (-1)^n A_2 + A_3 n + 3A_3.$$

Thus
$$y_{n+2} - y_n = 2A_3$$
$$y_{n+3} - y_{n+1} = 2A_3,$$

i.e.
$$y_{n+2} - y_n = y_{n+3} - y_{n+1},$$

a difference equation of the third order.

EXERCISES (III)

1. Eliminate the constants from the following expressions and hence in each case derive the corresponding difference equation of the lowest order possible and verify by direct insertion that the expression satisfies it.

(i) $y_n = A2^n + B3^n$
(ii) $y_n = (A + Bn)4^n$
(iii) $2^n = A^{y_n}$
(iv) $y_n = \cos\alpha + n\sin\alpha.$
(v) $y_n = \cos\alpha + n\sin\beta.$

2. Show that $y_n = An + f(A)$ satisfies the difference equation
$$f(\Delta y_n) + n\Delta y_n - y_n = 0,$$
A being an arbitrary constant.

Theorems

1. A difference equation of the rth order, viz.

$$F(n, y_n, y_{n+1}, \ldots, y_{n+r}) = 0$$

associated with r independent subsidiary conditions which enable r definite values of y to be specified, has a unique solution.

Let us suppose that the difference equation has been algebraically solved for y_{n+r}. Then each expression of the form

$$y_{n+r} = G(n, y_n, \ldots, y_{n+r-1})$$

is a recurrence relation. Putting $n = 0$, y_r can be calculated if $y_0, y_1, \ldots, y_{r-1}$ are known. Once this has been done, $y_{r+1}, y_{r+2}, \ldots$ can be found in succession by merely taking $n = 1, 2, 3$ etc.

Thus, y_{n+r} is fixed uniquely if r values of y at the successive unit intervals $n = 0, 1, 2$ etc. are specified. This is equivalent to the statement that if these r values of y are arbitrarily selected the solution of the difference equation contains r arbitrary constants.

The given values of y need not in fact occur for successive values of n; this point is brought out in the examples which follow.

2. If the solution of the rth order difference equation

$$F(n, y_n, y_{n+1}, \ldots, y_{n+r}) = 0$$

is known, then the solution of the difference equation

$$F(n, z_{n+s}, z_{n+s+1}, \ldots, z_{n+r+s}) = 0$$

is also known. This is obvious by writing $z_{n+s} = y_n$ in the original equation. Furthermore, by writing $z_{n-s} = y_n$, the solution of

$$F(n, z_{n-s}, z_{n-s+1}, \ldots, z_{n-s+r}) = 0$$

can be obtained.

Examples

1. To solve the equation

$$(n + 1)z_{n+5} - nz_{n+4} = 0.$$

Write

$$z_{n+4} = y_n,$$

or

$$z_n = y_{n-4},$$

thus the equation becomes

$$(n + 1)y_{n+1} - ny_n = 0,$$

and it can easily be shown that the solution of this equation is

$$y_n = A/n,$$

where A is an arbitrary constant. Thus

$$z_n = y_{n-4} = A/(n - 4).$$

2. The solution of

$$y_{n+1} = (n + 1)y_n$$

is

$$y_n = An!$$

where A is an arbitrary constant. Hence the solution of

$$u_{n+1} = (n + s + 1)u_n$$

is

$$u_n = A(n + s)!$$

3. If the solution of the difference equation

$$F(n, y_n, y_{n+1}, \ldots, y_{n+r}) = 0$$

is known, then the solution of the difference equation

$$F[n, f(z_n), f(z_{n+1}), \ldots, f(z_{n+r})] = 0$$

is also known. The latter equation is immediately transformed to the former by writing

$$y_n = f(z_n)$$

and thus

$$y_{n+1} = f(z_{n+1})$$

and

$$y_{n+2} = f(z_{n+2})$$

$$\cdots\cdots\cdots$$

$$y_{n+r} = f(z_{n+r}).$$

4. The solution of

$$y_{n+1} - y_n = 1$$

is

$$y_n = n + A,$$

where A is an arbitrary constant. Let

$$y_n = z_n^2,$$

thus

$$y_{n+1} = z_{n+1}^2,$$

then the solution of

$$z_{n+1}^2 - z_n^2 = 1$$

is

$$z_n^2 = n + A$$

or

$$z_n = \pm \sqrt{(n + A)}.$$

EXERCISES (IV)

Solve the following difference equations, using the solutions already given in earlier cases—

1. $y_{n+2}^2 - 5y_{n+1}^2 + 6y_n^2 = 0.$
2. $(n + 1)y^3_{n+1} - ny_n^3 = 0.$
3. $(n + 1)^2 y_{n+1} - n^2 y_n = 0.$
4. $\sqrt{y_{n+1}} = n\sqrt{y_n}.$
5. Verify that $y_n = A\lambda^n$ satisfies the equation

$$y_{n+1} = \lambda y_n,$$

and that

$$y_n = A\lambda^n + \mu/(1 - \lambda)$$

satisfies

$$y_{n+1} = \lambda y_n + \mu.$$

Hence solve the equations—

(i) $y_{n+1} = y_n^{\lambda}$,

(ii) $e^{\Delta y_n} = \lambda e^{y_n}$

Functional Equations or Difference Equations of a Continuous Independent Variable

In all the foregoing examples where we have illustrated how difference equations arise, and on occasion what their solutions are and how they are uniquely specified, the independent variable n has been restricted to be an integer. We propose now to extend this concept to the case where x, the continuous variable replaces n the integer. The problem then is to determine the function $y(x)$ from some relation, say, of the form

$$y(x+1) = F[x, y(x)] \tag{8}$$

We term this a functional equation of the first order, just as

$$y_{n+1} = F(n, y_n) \tag{9}$$

was termed a difference equation of the first order. In the same way

$$y(x+r) = F(x, y(x), y(x+1), \ldots, y(x+r-1)) \tag{10}$$

is a functional equation of the rth order.

There is, however, no hard and fast rule for the use of these terms. An equation of type (9) is certainly always referred to as a difference equation and an equation of type

$$y(2x) = F[x, y(x)],$$

say, in which the variable does not change by a step of one unit, is always referred to as a functional equation. An equation of type (8) or (9) on the other hand, which is a functional equation by our definition, is sometimes also referred to as a difference equation in virtue of the fact that the independent variable changes by step unity. The distinction we have drawn between (8) and (9) rests directly on the fact that in the case of (8) the independent variable adopts all values in a given range, whereas in the case of (9) only integral values enter. We shall therefore *always* refer to an equation of type (9) as a difference equation but we may refer to an equation of type (8) either as a difference equation or as a functional equation.

As we have seen, an equation of the form (9) arose from the relation

$$y_n = f(n, A)$$

where A was an arbitrary constant. Consider instead, a relation of the form

$$y(x) = f[x, A(x)] \tag{11}$$

where $A(x)$ is a periodic function of x of period unity. Such a function will in future be referred to as a ***unit periodic***. Increasing x to $x + 1$ in (11), and remembering that $A(x + 1) = A(x)$, we have

$$y(x + 1) = f[x + 1, A(x)]. \tag{12}$$

Accordingly we may now eliminate $A(x)$ between (11) and (12), and we derive the functional equation of the first order

$$y(x + 1) = F[x, y(x)]. \tag{13}$$

The expression (11) is then the solution of (13) and contains, not an arbitrary constant, but an arbitrary unit periodic. In the same way it is clear that

$$y(x) = f[x, A_1(x), A_2(x), \ldots, A_r(x)],$$

where $A_1(x), A_2(x), \ldots, A_r(x)$ are arbitrary unit periodics, is the solution of a functional equation of the form

$$y(x + r) = F[x, y(x), y(x + 1), \ldots, y(x + r - 1)].$$

Expression for a Unit Periodic

A function $A(x)$ with unit periodicity can be expressed in general by Fourier's theorem in the form

$$A(x) = a_0 + a_1 \cos 2\pi x + \ldots + a_n \cos 2n\pi x + \ldots$$
$$+ b_1 \sin 2\pi x + \ldots + b_n \sin 2n\pi x + \ldots,$$

where $a_0, a_1, \ldots, a_n, \ldots, b_1, b_2, \ldots, b_n, \ldots$ are constants. If, for x positive, we write $x = N + [x]$ where N is the greatest integer in x and $[x]$ is the fractional part, then any function of $[x]$, say $f([x])$ will be a unit periodic which, over the range $0 \leqslant x < 1$ will coincide with $f(x)$. This, for many purposes we have in mind, will be a more convenient method of representing the unit periodic than by expressing it as a Fourier series.

We have seen that when dealing with a difference equation in y_n the presence of an arbitrary constant in the solution corresponds to the specifying of a particular value, y_1, of y_n. When the difference equation is one in $y(x)$ the arbitrary constant is replaced by an arbitrary periodic. What in that case corresponds to the specification of a particular value of the dependent variable y?

Consider the first order equation

$$y(x + 1) = F[x, y(x)] \tag{14}$$

If $y(x)$ is known over the range $0 \leqslant x < 1$ then, from (14), $y(x + 1)$ may be determined over the same range of x, i.e. $y(x)$ is determinate over the range $1 \leqslant x < 2$. By inserting this on the right $y(x)$ may then be determined over the range $2 \leqslant x < 3$, and so on. Thus, given $y(x)$ over the unit range $0 \leqslant x < 1$, equation (14) fixes the function $y(x)$ for all positive values of x. Furthermore, by solving (14) algebraically for $y(x)$ in terms of x and $y(x + 1)$ we may proceed to find $y(x)$ over all negative values of x. We conclude that the solution of (14) will be uniquely determinate if $y(x)$ is given over the range $0 \leqslant x < 1$.

It may happen of course, that a solution of the functional equation possesses some form of symmetry which enables $y(x)$ to be related over part of the range $0 \leqslant x < 1$ to the remainder of that range. In such circumstances $y(x)$ would require to be given over only one part of the range in order that it may be completely determined. For example, we shall see later that a certain solution of the difference equation

$$y(x + 1) = xy(x)$$

where x may adopt any value between $-\infty$ and $+\infty$ implies the relation

$$y(x)y(1 - x) = \pi/\sin \pi x.$$

From this it will follow that the complete solution of the original equation requires $y(x)$ to be specified over one half of the range $0 - 1$ only, say $0 \leqslant x < \frac{1}{2}$.

However, the fact that the functional equation (14) enables us to calculate $y(x)$ step-by-step in this way does not in itself help us much in any examination of the properties of $y(x)$. It is the formal solution of (14) to which we must turn in general for this purpose, and as we have seen, this formal solution, that is, an expression for y explicitly in terms of x, may include a function of x of period unity. It is obvious from what we have said that this periodic function must be uniquely determined from the given form of $y(x)$ over the range $0 \leqslant x < 1$. The solution of equation (14) then may be written

$$F[x, y(x), A(x)] = 0 \tag{15}$$

where $A(x)$ is the unit periodic. Thus we may write

$$A(x) = Q[x, y(x)].$$

If $Q[x, y(x)]$ has a finite number of poles and discontinuities in the range $0 \leqslant x < 1$, $y(x)$ being given in that range, it is possible to expand $A(x)$ as a Fourier series of the form

$$a_0 + a_1 \cos 2\pi x + a_2 \cos 4\pi x + a_3 \cos 6\pi x + \ldots$$
$$+ b_1 \sin 2\pi x + b_2 \sin 4\pi x + \ldots$$

and so the function $A(x)$ in (15) is uniquely determined.

Example

Solve $y(x + 1) = ey(x)$ where we are given that $y(x) = xe^x$ over the range $0 \leqslant x < 1$. We may write $y(x) = A(x)e^x$ where $A(x)$ is a unit periodic. This satisfies the equation. Accordingly over the range $0 \leqslant x < 1$, $A(x) = x$. Writing $x = [x] + N$, where N is the integral part of x, and $[x]$ its fractional part for any value of x, then $[x]$ is itself a periodic function of period unity. Thus the solution of the functional equation is $y(x) = [x]e^x$. Actually $[x]$ may be expressed in terms of its Fourier expansion. Thus—

$$[x] = \frac{1}{2} - \frac{1}{\pi}\left[\frac{\sin 2\pi x}{1} + \frac{\sin 4\pi x}{2} + \frac{\sin 6\pi x}{3} + \ldots\right].$$

This procedure may quite obviously be generalized to an equation of the rth order, the generalization being clear from the following illustration.

Consider the functional equation—

$$y(x + 2) = F[y(x + 1), y(x), x].$$

It is now obvious from this form that if $y(x)$ is specified over two intervals, say $0 \leqslant x < 1$ and $1 \leqslant x < 2$, then the functional equation enables us to calculate $y(x)$ everywhere. For reasons we have already explained we conclude that the solution of this final equation will involve two periodic functions. Thus

$$y(x) = G[x, A_1(x), A_2(x)],$$

and accordingly,

$$\begin{aligned} y(x + 1) &= G[x + 1, A_1(x + 1), A_2(x + 1)] \\ &= G[x + 1, A_1(x), A_2(x)]. \end{aligned}$$

Solving these two equations for $A_1(x)$ and $A_2(x)$, we obtain two expressions of the form—

$$A_1(x) = Q_1[x, y(x), y(x+1)],$$
$$A_2(x) = Q_2[x, y(x), y(x+1)].$$

Since $y(x)$ and $y(x+1)$ are given over a single period $0 \leqslant x < 1$, it follows that $A_1(x)$ and $A_2(x)$ are determinate over that period. As in the previous illustration, these may now be expanded in Fourier series. This then enables the final solution to be expressed in the form

$$y(x) = G[x, A_1(x), A_2(x)].$$

As we have stated, generalization to a difference equation of the rth order is immediate.

Example

If $$y(x+2) - 5y(x+1) + 6y(x) = 0,$$

then $$y(x) = A_1(x)2^x + A_2(x)3^x,$$

where $A_1(x)$ and $A_2(x)$ are unit periodics. Say $y(x) = x$ over the range $0 \leqslant x < 1$, and $y(x) = 1$, over the range $1 \leqslant x < 2$.

Now, $$y(x+1) = 2A_1(x)2^x + 3A_2(x)3^x,$$

hence $$y(x+1) - 2y(x) = A_2(x)3^x$$

or $$A_2(x) = 3^{-x}\{y(x+1) - 2y(x)\},$$

and $$3y(x) - y(x+1) = A_1(x)2^x$$

or $$A_1(x) = 2^{-x}\{3y(x) - y(x+1)\}.$$

Inserting the boundary conditions, we have—

$$A_2(x) = 3^{-[x]}(1 - 2[x]),$$

and $$A_1(x) = 2^{-[x]}(3[x] - 1).$$

Thus finally, for x positive,

$$y(x) = 2^N(3[x] - 1) + 3^N(1 - 2[x])$$

where $x = [x] + N$ and N is the greatest integer in x.

CHAPTER 4

Linear Difference and Functional Equations with Constant Coefficients

We proceed to the complete solution of a certain class of difference equation analogous to the linear differential equation with constant coefficients.

When the terms of a difference equation involve y_n, y_{n+1}, y_{n+2}, . . ., y_{n+r} to the first degree we say that the equation is linear and of the rth order. We begin with a few simple examples developing gradually in generality.

1. Consider the linear difference or functional equation of the first order—

$$y(x + 1) - y(x) = 0,$$

i.e.

$$\Delta y(x) = 0.$$

This asserts that the successive values of y at intervals of unity in x are all equal. After unit interval in x the values of y will repeat. This means in effect that $y(x)$ is a periodic function of x of period unity. Such a function we have called a unit periodic.

If x be replaced by n, an integer, then of course the difference equation

$$y_{n+1} - y_n = 0$$

implies that y_n is constant.

2. The equation

$$y(x + 1) - \lambda y(x) = 0,$$

where λ is a given constant can be easily reduced to the above form for we may write the equation

$$y(x + 1)/\lambda^{x+1} - y(x)/\lambda^x = 0,$$

or

$$\Delta[y(x)/\lambda^x] = 0.$$

Accordingly,

$$y(x)/\lambda^x = A(x)$$

where $A(x)$ is an arbitrary unit periodic. Thus

$$y(x) = A(x)\lambda^x$$

is the solution of the first order linear difference equation with constant coefficients, viz.

$$y(x+1) - \lambda y(x) = 0,$$

or, as we may equally write it

$$(\mathrm{E} - \lambda)y(x) = 0.$$

This simple example now enables us to deal with equations of higher order than the first.

3. Consider therefore the linear equation of the second order with constant coefficients—

$$y(x+2) + 2\lambda y(x+1) + \mu y(x) = 0.$$

We may write this in the operational form

$$(\mathrm{E}^2 + 2\lambda \mathrm{E} + \mu)y(x) = 0$$

or $$(\mathrm{E} - \alpha)(\mathrm{E} - \beta)y(x) = 0;$$

or equally, $$(\mathrm{E} - \beta)(\mathrm{E} - \alpha)y(x) = 0.$$

Here α and β are the roots of the equation

$$z^2 + 2\lambda z + \mu = 0.$$

Now if $y(x)$ be a function of x which satisfies the subsidiary equation

$$(\mathrm{E} - \alpha)y(x) = 0,$$

then it will also satisfy the equation

$$(\mathrm{E} - \beta)(\mathrm{E} - \alpha)y(x) = 0.$$

In the same way if $y(x)$ be a function of x which satisfies the subsidiary equation

$$(\mathrm{E} - \beta)y(x) = 0,$$

then it will also satisfy the equation

$$(\mathrm{E} - \alpha)(\mathrm{E} - \beta)y(x) = 0.$$

It follows that we can derive two independent solutions of the original equation by solving the two *subsidiary* equations—

$$(\mathrm{E} - \alpha)y(x) = 0 \qquad \text{and} \qquad (\mathrm{E} - \beta)y(x) = 0,$$

respectively, giving

$$y(x) = A\alpha^x \quad \text{and} \quad y(x) = B\beta^x,$$

where A and B are arbitrary unit periodics. It is easily verified that the expression $A\alpha^x + B\beta^x$ satisfies the original equation, and that since it contains two arbitrary periodics it is also the general solution. The generalization of this to equations of order higher than two is obvious.

Examples

1. $$y(x+2) - 7y(x+1) + 10y(x) = 0,$$

i.e. $$(\mathrm{E}^2 - 7\mathrm{E} + 10)y(x) = 0,$$

$$(\mathrm{E} - 2)(\mathrm{E} - 5)y(x) = 0.$$

Thus $$y(x) = A\,.\,2^x + B\,.\,5^x,$$

where A and B are arbitrary unit periodics.

2. $$y(x+2) + 7y(x+1) + 10y(x) = 0,$$

i.e. $$(\mathrm{E} + 2)(\mathrm{E} + 5)y(x) = 0.$$

Thus $$y(x) = A(-2)^x + B(-5)^x,$$

where, as usual, A and B are arbitrary unit periodics.

3. $$y(x+3) - 6y(x+2) + 11y(x+1) - 6y(x) = 0,$$

i.e. $$(\mathrm{E}^3 - 6\mathrm{E}^2 + 11\mathrm{E} - 6)y(x) = 0,$$

or $$(\mathrm{E} - 3)(\mathrm{E} - 1)(\mathrm{E} - 2)y(x) = 0.$$

Thus $$y(x) = A2^x + B3^x + C.$$

When the operational factors are complex, i.e. when $(\lambda^2 - \mu) < 0$, it is advantageous to write the solution in a different form. In this case,

$$\alpha = a + ib \quad \text{and} \quad \beta = a - ib$$

where $$a = -\lambda \quad \text{and} \quad b = +\sqrt{(\mu - \lambda^2)},$$

so that

$$y(x) = A\alpha^x + B\beta^x = A(a + ib)^x + B(a - ib)^x$$
$$= (a^2 + b^2)^{\frac{1}{2}x}\{A(\cos x\theta + \mathrm{i}\sin x\theta) + B(\cos x\theta - \mathrm{i}\sin x\theta)\}$$

where $\tan\theta = b/a,$

or
$$y(x) = \mu^{\frac{1}{2}x}\{(A + B)\cos x\theta + 1(A - B)\sin x\theta\}.$$

Thus
$$y(x) = C\mu^{\frac{1}{2}x}\cos(x\theta + G),$$

where C and G are arbitrary unit periodics and
$$\tan\theta = -(\mu - \lambda^2)^{\frac{1}{2}}/\lambda$$
is the solution of the difference equation
$$y(x + 2) + 2\lambda y(x + 1) + \mu y(x) = 0,$$
in a form suitable for the case when $\lambda^2 < \mu$.

Example
$$y(x + 2) - 2y(x + 1) + 5y(x) = 0,$$
$$(\mathrm{E}^2 - 2\mathrm{E} + 5)y(x) = 0,$$
$$[\mathrm{E} - (1 + 2\mathrm{i})][\mathrm{E} - (1 - 2\mathrm{i})]y(x) = 0.$$

Hence
$$y(x) = A(1 + 2\mathrm{i})^x + B(1 - 2\mathrm{i})^x$$
$$= C5^{\frac{1}{2}x}\cos(x\theta + G),$$

where C and G are arbitrary unit periodics and $\theta = \tan^{-1} 2$.

Case where the Operational Factors are Identical

Here the equation takes the form
$$y(x + 2) - 2\lambda y(x + 1) + \lambda^2 y(x) = 0,$$
or
$$(\mathrm{E} - \lambda)^2 y(x) = 0.$$

Let $y(x) = \lambda^x z(x)$ where $z(x)$ is a new dependent variable. Then inserting this into the difference equation, we obtain—
$$\lambda^{x+2}z(x + 2) - 2\lambda^{x+2}z(x + 1) + \lambda^{x+2}z(x) = 0.$$

Thus,
$$z(x + 2) - 2z(x + 1) + z(x) = 0,$$
or
$$\Delta^2 z(x) = 0;$$
hence
$$z(x) = Ax + B.$$

Accordingly the solution of the equation
$$(\mathrm{E} - \lambda)^2 y(x) = 0$$
is
$$y(x) = \lambda^x(Ax + B)$$
where A and B are arbitrary periodic functions of x of period unity.

In the case where x can take integral values only, A and B become arbitrary constants.

Examples

1. To find y_n if $y_{n+2} - 6y_{n+1} + 9y_n = 0$ and $y_0 = 1$ and $y_1 = 0$.

Here $$(E^2 - 6E + 9)y_n = 0,$$

So that $$(E - 3)^2 y_n = 0$$

Hence $$y_n = (An + B)3^n$$

But $y_0 = 1$ and $y = 0$, so $1 = B$ and $0 = 3(A + B)$. Thus

$$y_n = (1 - n) . 3^n.$$

2. Evaluation of the determinant

$$D_n = \begin{vmatrix} \alpha & 1 & 0 & 0 & \cdots \\ 1 & \alpha & 1 & 0 & \cdots \\ 0 & 1 & \alpha & 1 & \cdots \\ 0 & 0 & 1 & \alpha & \cdots \\ \cdot & \cdot & \cdot & \cdot & \cdots \\ \cdot & & & & \\ \cdot & & & & \end{vmatrix}$$

where D_n has n rows and n columns.

By direct expansion it is clear that

$$D_n = \alpha D_{n-1} - D_{n-2}$$

or $$D_{n+2} - \alpha D_{n+1} + D_n = 0 \text{ by changing } n \text{ to } n + 2,$$

where $D_1 = \alpha$, $D_2 = \alpha^2 - 1$.

Hence $$(E^2 - \alpha E + 1)D_n = 0$$

or $$\left[E - \frac{\alpha + i\sqrt{(4 - \alpha^2)}}{2}\right]\left[E - \frac{\alpha - i\sqrt{(4 - \alpha^2)}}{2}\right] D_n = 0$$

or $$[E - (\cos\theta + i\sin\theta)][E - (\cos\theta - i\sin\theta)]D_n = 0$$

where $\tan\theta = \sqrt{(4 - \alpha^2)}/\alpha$, assuming $|\alpha| < 2$. Hence

$$\begin{aligned} D_n &= A(\cos\theta + i\sin\theta)^n + B(\cos\theta - i\sin\theta)^n \\ &= A(\cos n\theta + i\sin n\theta) + B(\cos n\theta - i\sin n\theta) \\ &= C\cos(n\theta + \beta) \end{aligned}$$

where C and β are arbitrary. Thus

$$\alpha = D_1 = C\cos(\theta + \beta)$$
$$\alpha^2 - 1 = D_2 = C\cos(2\theta + \beta)$$

giving $D_n = \sin (n + 1)\theta/\sin \theta$

$$= 2 \sin [(n + 1) \tan^{-1}\{\sqrt{(4 - \alpha^2)}/\alpha\}]/\sqrt{(4 - \alpha^2)}.$$

EXERCISES (I)

1. Evaluate D_n when $|\alpha| > 2$.
2. Show that when $\alpha = 2$, $D_n = n + 1$.

Example

The successive terms $f(n)$, $f(n + 1)$, $f(n + 2)$ of a sequence are always connected by a relation of the form

$$f(n + 2) + \lambda f(n + 1) + \mu f(n) = 0$$

where λ and μ are constants. The values of $f(0)$, $f(1)$, $f(2)$, and $f(3)$ are respectively 0, 1, 2, 2. Determine the expression for $f(n)$.

Using the relation above for these integral values of n we have

$$2 + 2\lambda + \mu = 0$$
$$2 + \lambda = 0$$

giving $$\lambda = -2 \qquad \mu = 2.$$

Thus $$f(n + 2) - 2f(n + 1) + 2f(n) = 0$$

or $$(\mathrm{E}^2 - 2\mathrm{E} + 2)f(n) = 0,$$

i.e. $$[\mathrm{E} - (1 + \mathrm{i})][\mathrm{E} - (1 - \mathrm{i})]f(n) = 0.$$

Accordingly $$f(n) = A(1 + \mathrm{i})^n + B(1 + \mathrm{i})^n.$$

Inserting the values for $f(0)$ and $f(1)$

$$0 = A + B$$
$$1 = A(1 + i) + B(1 - i) = i(A - B).$$

Hence $$2A = -\mathrm{i}, \qquad 2B = \mathrm{i}$$

and $$2f(n) = \mathrm{i}(1 - \mathrm{i})^n - \mathrm{i}(1 + \mathrm{i})^n.$$

But $$(1 + \mathrm{i})^n = e^{n \log (1+\mathrm{i})} = e^{n \log \sqrt{2}} \,.\, e^{\mathrm{i}n \tan^{-1} 1}$$

$$= 2^{n/2}\left(\cos \frac{n\pi}{4} + \mathrm{i} \sin \frac{n\pi}{4}\right).$$

It follows that

$$2f(n) = \mathrm{i}2^{n/2}\left(\cos \frac{n\pi}{4} - \mathrm{i} \sin \frac{n\pi}{4}\right) - \mathrm{i}2^{n/2}\left(\cos \frac{n\pi}{4} + \mathrm{i} \sin \frac{n\pi}{4}\right)$$

or $f(n) = 2^{n/2} \sin n\pi/4$.

Check

$$n = 3 \qquad f(3) = 2^{3/2} \sin \frac{3\pi}{4} = 2^{3/2} \times 2^{-1/2} = 2$$

and

$$f(n + 2) - 2f(n + 1) + 2f(n)$$

$$= 2 \,.\, 2^{n/2} \sin\left(\frac{n\pi}{4} + \frac{\pi}{2}\right) - 2 \,.\, 2^{1/2} \,.\, 2^{n/2} \sin\left(\frac{n\pi}{4} + \frac{\pi}{4}\right)$$

$$+ 2 \,.\, 2^{n/2} \sin \frac{n\pi}{4} = 0.$$

EXERCISES (II)

1. Solve the following functional equations for $y(x)$—
   (i) $y(x + 2) - y(x + 1) - 2y(x) = 0$,
   (ii) $y(x + 2) - 5y(x + 1) + 4y(x) = 0$,
   (iii) $y(x + 2) + y(x) = 0$,
   (iv) $y(x + 2) - 4y(x + 1) + 5y(x) = 0$,
   (v) $y(x + 2) - 6y(x + 1) + 9y(x) = 0$,
   (vi) $y(x + 2) - 4y(x + 1) + 4y(x) = 0$.

2. A series of values of y_n satisfy the relation

$$y_{n+2} + 2\lambda y_{n+1} + \mu y_n = 0.$$

Given that $y_0 = 0$, $y_1 = 3$, $y_2 = 6$, $y_3 = 36$, find y_n.

3. Three successive values of the sequences y_n satisfy the relation

$$y_n \cdot y_{n+2} = y_{n+1}^2.$$

Determine the expression for y_n.

4. Show that if $y(x + 2) - 2\lambda y(x + 1) + \lambda^2 y(x) = 0$, then

$$\lim_{x\to\infty} [y(x + 1)/y(x)] = \lambda$$

irrespective of the boundary conditions.

5. Show that $16y(x + 2) - 8y(x + 1) + y(x)$ can be converted into a perfect difference by multiplying by 2^{2x}. Hence solve—

$$16y(x + 2) - 8y(x + 1) + y(x) = 0.$$

6. Determine the coefficient of x^n in the expansions of the following functions in ascending powers of x—
   (i) $(3x - 4)/(x^2 + 5x + 4)$,
   (ii) $(x - 1)/(4x^2 + 4x + 1)$.

The General Linear Difference Equation—Elementary Properties of its Solution

Certain properties of the linear difference equation are analogous to those of the linear differential equation. Restricting ourselves for the moment to real integral values of the independent variable the difference equation may be written

$$y_{n+r} + p_1 y_{n+r-1} + p_2 y_{n+r-2} + \cdots + p_r y_n = f(n) \tag{1}$$

where p_1, p_2, . . . may be functions of n.

1. If y_n is given at r successive values of n then equation (1) enables us to extend this to $(r + 1)$ successive values. Then by raising n to $(n + 1)$ this can be extended one stage further and so on indefinitely. This means that r successive values of y_n will suffice for the solution of (1). It follows that the general solution of (1) will contain r arbitrary constants and hence the r given values need not be successive. This is simply a particular case of the general property to which we referred earlier.

2. If Y_n and Z_n are two independent solutions of the equation

$$y_{n+r} + p_1 y_{n+r-1} + p_2 y_{n+r-2} + \cdots + p_r y_n = 0 \tag{2}$$

then $A \,.\, Y_n + B \,.\, Z_n$, where A and B are arbitrary constants, is also a solution. This is obvious at once by direct insertion into the equation. It follows that if r such solutions are found and each is multiplied by an arbitrary constant and their sum be taken, this sum is not only *a* solution but is the most general solution of equation (2). This expression we call the *complementary function* of the solution of (1).

3. If P_n is a *particular solution* of (1), i.e. *any* function of n which satisfies (1), then the general solution of (1) is the general solution of (2), i.e. the complementary function of the solution of (1), plus this *particular solution*. This follows from the fact that the sum of the complementary function and the particular solution satisfies equation (1) and contains the appropriate number of arbitrary constants.

We now remark that what we have said with regard to the general form of the solution of (1) is equally valid for the more general case where the independent variable is not restricted to n, an integer, but is the continuous variable x, with this difference, that in place of the r arbitrary constants we now have r arbitrary unit periodics. The boundary conditions that will then be necessary to fix the solution uniquely will be more complicated. We shall examine this in greater detail later. For the moment familiarity with the processes will be derived by the more restricted study of cases where the variable is n. This will then pave the way for the more general treatment.

It remains only to show how the complementary function and the particular solutions are to be derived. Consider

$$(\mathrm{E} - \alpha)^p y(x) = 0.$$

Let $$y(x) = \alpha^x z(x).$$

Then
$$\begin{aligned}(\mathrm{E} - \alpha)y(x) &= (\mathrm{E} - \alpha)\alpha^x z(x)\\ &= \alpha^{x+1}z(x+1) - \alpha^{x+1}\,.\,z(x)\\ &= \alpha\,.\,\alpha^x \Delta z(x).\end{aligned}$$

If we operate again with $\mathrm{E} - \alpha$ the right-hand side is of the same form as before, viz. $\alpha^x z(x)$. Hence

$$(\mathrm{E} - \alpha)^2 y(x) = \alpha^2 \alpha^x \Delta^2 z(x);$$

and generally, $$(\mathrm{E} - \alpha)^p y(x) = \alpha^p \alpha^x \Delta^p z(x).$$

Accordingly the solution of the equation

$$(\mathrm{E} - \alpha)^p y(x) = 0,$$

where $$y(x) = \alpha^x z(x),$$

is found from the solution of

$$\Delta^p z(x) = 0.$$

Thus $z(x)$ is a polynomial in x of degree $p - 1$ and consequently—

$$y(x) = \alpha^x[A_0 + A_1 x + A_2 x(x-1) + \ldots + A_{p-1}x(x-1)\ldots(x-p+1)],$$

an expression that contains the necessary number of arbitrary unit periodics. When in place of x we have the integer n, these arbitrary periodics reduce to arbitrary constants.

The general solution therefore of the linear equation of any order is derived at once by adding together all the general solutions of the subsidiary equations, as found by the above method.

Examples

1. $$y_{n+3} - 7y_{n+2} + 16y_{n+1} - 12y_n = 0.$$

Thus $$(\mathrm{E}^3 - 7\mathrm{E}^2 + 16\mathrm{E} - 12)y_n = 0,$$

or $$(\mathrm{E} - 2)^2(\mathrm{E} - 3)y_n = 0.$$

The general solution is therefore

$$y_n = A\,.\,3^n + (Bn + C)2^n,$$

which can be verified by direct substitution. In order that the terms in the sequence y_n might be calculable from the original equation, we require y_1, y_2, and y_3 to be known. These values, or any three values of y_n, therefore, should determine A, B, and C. Suppose $y_1 = -1$, $y_2 = -3$, $y_3 = -5$, then from the general solution we have

$$-1 = 3A + 2(B + C),$$
$$-3 = 9A + 4(2B + C),$$
$$-5 = 27A + 8(3B + C),$$

and these give $A = 1$, $B = C = -1$. Hence

$$y_n = 3^n - (n + 1)2^n.$$

2. $$y_{n+4} - 2y_{n+3} + 2y_{n+2} - 2y_{n+1} + y_n = 0$$

where $y_1 = 1$, $y_2 = 0$, $y_3 = 1$, $y_5 = 1$.

$$(\mathrm{E}^4 - 2\mathrm{E}^3 + 2\mathrm{E}^2 - 2\mathrm{E} + 1)y_n = 0,$$
$$(\mathrm{E} - 1)^2(\mathrm{E}^2 + 1)y_n = 0.$$

Thus
$$y_n = (A + Bn)(1)^n + C\mathrm{i}^n + D(-\mathrm{i})^n,$$
$$= A + Bn + C\,\mathrm{e}^{n\mathrm{i}\pi/2} + D\,\mathrm{e}^{-n\mathrm{i}\pi/2},$$
$$= A + Bn + P\cos\frac{n\pi}{2} + Q\sin\frac{n\pi}{2}$$

where A, B, P, and Q are arbitrary constants. Inserting the boundary conditions, we have

$$1 = y_1 = A + B + Q,$$
$$0 = y_2 = A + 2B - P,$$
$$1 = y_3 = A + 3B - Q,$$
$$1 = y_5 = A + 5B + Q,$$

from which it follows that

$$A = 1, \quad B = 0, \quad P = 1, \quad Q = 0, \quad y_n = 1 + \cos\frac{n\pi}{2}.$$

Pseudo-non-linear Equations

Non-linear difference equations can sometimes be transformed either through a change of dependent or of independent variable into a linear difference equation. In this chapter we shall restrict ourselves to transformations in the dependent variable, viz. y, and not in the independent variable x or n. It will be apparent how much easier it is in the case of difference equations to carry through such transformations than in the case of differential equations. On the other hand, it is much more difficult to effect changes in the independent variable.

Examples

1. $$y^2(x+1) - y^2(x) = 1.$$

Here we need only write

$$z(x) = y^2(x),$$

and the equation becomes

$$z(x+1) - z(x) = 1.$$

Hence
$$z(x) = x + A,$$

where A is an arbitrary unit periodic. Hence

$$y^2(x) = x + A,$$

is the general solution of the difference equation.

2. $$y_{n+1} = \sqrt{\{y_n + \sqrt{[y_{n-1} + \sqrt{(y_{n-2} + \cdots)}]}\}}.$$

Hence
$$\begin{aligned} y_{n+1}^2 &= y_n + \sqrt{[y_{n-1} + \sqrt{y_{n-2}} + \cdots]} \\ &= y_n + y_n = 2y_n. \end{aligned}$$

Taking logarithms of both sides, and writing $\log y_n = z_n$, we obtain—

$$2z_{n+1} - z_n = \log 2.$$

To find a particular solution of this equation, try $z_n = c$, a constant, then

$$2c - c = \log 2.$$

Thus the general solution of the equation is

$$z_n = A \,.\, (1/2)^n + \log 2.$$

Hence
$$y_n = 2B^{(1/2)^n}$$

where B is an arbitrary constant. From the original equation

$$y_2 = \sqrt{y_1}, \qquad y_3 = \sqrt{[y_2 + \sqrt{y_1}]} = \sqrt{(2y_2)}.$$

If $y_1 = 1$ then $y_2 = 1$, $y_3 = \sqrt{2}$. But from the solution we have $y_1 = 2B^{1/2}$, i.e. $B = 1/4$, and $y_n = 2^{(1-1/2^{n-1})}$.

Check

$$y_2 = 2^{1-1/2} = \sqrt{2}.$$

EXERCISES (III)

Solve the equations—

1. $y_n \cdot y_{n+2} = y^2_{n+1}$.
2. $y_{n+3}^2 - 3y_{n+2}^2 + 3y_{n+1}^2 - y_n^2 = 0.$
3. $y_{n+1} \cdot y_n + 4y_{n+2} \cdot y_n - 21y_{n+2} \cdot y_{n+1} = e^n y_{n+1} \cdot y_{n+2} \cdot y_n$.
4. $\Delta^2 y_n = (\Delta y_n)(\Delta y_{n+1})$.
5. $n(\Delta y_n)(\Delta y_{n+1})(\Delta y_{n+2}) = (\Delta y_{n+2})(\Delta^2 y_n) - (\Delta y_n)(\Delta^2 y_{n+1})$.
6. $y_{n+4} \cdot y_{n+2}^6 y_n = y_{n+1}^4 \cdot y_{n+3}^4$.

3. Four successive ordinates of a curve y_n, y_{n+1}, y_{n+2}, y_{n+3} are related in such a way that

$$y_n y_{n+2}^3 = y_{n+1}^3 \cdot y_{n+3}.$$

If $y_0 = 1/\sqrt{(2\pi)}$, $y_1 = 1/\sqrt{(2\pi\, e)}$, $y_2 = 1/\sqrt{(2\pi\, e^4)}$, determine the expression for y_n. Here

$$\log y_n + 3 \log y_{n+2} = 3 \log y_{n+1} + \log y_{n+3}.$$

Write $u_n = \log y_n$, and it follows that

$$u_{n+3} - 3u_{n+2} + 3u_{n+1} - u_n = 0,$$

or

$$(E - 1)^3 u_n = 0.$$

Hence

$$u_n = A + Bn + Cn^2$$

or

$$y_n = e^{u_n} = \lambda\, e^{Bn+Cn^2}.$$

Thus

$$y_0 = 1/\sqrt{(2\pi)} = \lambda,$$

$$y_1 = 1/\sqrt{(2\pi\, e)} = \lambda\, e^{B+C}; \quad \text{or} \quad e^{B+C} = 1/\sqrt{e}.$$

Again $y_2 = 1/\sqrt{(2\pi\, e^4)} = \lambda\, e^{2B+4C}$; or $e^{2B+4C} = 1/e^2$.

Hence

$$B + C = -1/2; \qquad B + 2C = -1,$$

i.e.

$$B = 0, \qquad C = -1/2,$$

and

$$y_n = \frac{1}{\sqrt{(2\pi)}}\, e^{-n^2/2}.$$

So far, except in special cases, we have restricted ourselves to solving linear difference equations in which the right-hand side is zero. Now the complete solution of a difference equation of linear form is the sum of the complementary function, which contains the appropriate number of arbitrary constants (or unit periodics), and a particular solution. We have therefore just seen how to determine the complementary function, and it remains to examine how to determine the particular solution. We shall not deal with this problem in its most general form, but restrict ourselves to the cases where the right-hand side is one of a set of special types of function.

Determination of the Particular Solution

Write the equation in the form

$$\phi(\mathrm{E})y(x) = f(x)$$

then the particular solution may be written symbolically as

$$P(x) = [\phi(\mathrm{E})]^{-1}f(x).$$

1. If $f(x)$ is a polynomial in x of finite degree r then we may expand $[\phi(\mathrm{E})]^{-1} = [\phi(1 + \Delta)]^{-1}$ in ascending powers of Δ, and operate with each of these terms on $f(x)$. Clearly we need carry the expansion only as far as Δ^r.

Example

$$y(x + 2) - 4y(x) = x^2 - 1,$$
$$(\mathrm{E}^2 - 4)y(x) = x^2 - 1.$$

Complementary function is $A\, 2^x + B(-2)^x$ where A and B are unit periodics.

Particular solution is—

$$P(x) = \frac{1}{\mathrm{E}^2 - 4}(x^2 - 1) = \frac{1}{4}\left(\frac{1}{\mathrm{E} - 2} - \frac{1}{\mathrm{E} + 2}\right)(x^2 - 1)$$
$$= \frac{1}{4}\left[-\frac{1}{1 - \Delta} - \frac{1}{3(1 + \Delta/3)}\right](x^2 - 1)$$
$$= \frac{1}{4}\left[-1 - \Delta - \Delta^2 \ldots - \frac{1}{3} + \frac{\Delta}{9} - \frac{\Delta^2}{27} \ldots\right](x^2 - 1)$$
$$= \frac{1}{4}\left[-\frac{4}{3} - \frac{8}{9}\Delta - \frac{28}{27}\Delta^2 + \ldots\right](x^2 - 1)$$

$$= \frac{1}{4}\left[-\frac{4}{3}(x^2 - 1) - \frac{8}{9}(2x + 1) - \frac{28}{27} \times 2\right]$$

$$= -\frac{x^2}{3} - \frac{4x}{9} - \frac{11}{27}.$$

2. If $f(x)$ is of the form $a^x F(x)$, $F(x)$ being a polynomial of finite degree in x, the equation is—

$$\phi(\mathrm{E}) \, . \, y(x) = a^x F(x),$$

where $\phi(\mathrm{E})$ is a polynomial function of E. Let

$$y(x) = a^x u(x)$$

then $$\phi(\mathrm{E})y(x) = \phi(\mathrm{E})a^x u(x) = a^x \phi(a\mathrm{E})u(x)$$

as already proved.

So we have reduced the equation in $y(x)$ to one in the new variable $u(x)$, viz.

$$\phi(a\mathrm{E})u(x) = F(x),$$

with a polynomial function on the right-hand side which is of the same type as in case 1.

When $F(x) = 1$, a particular solution is

$$y(x) = [1/\phi(\mathrm{E})]a^x = a^x/\phi(a)$$

provided $\phi(a) \neq 0$.

3. A particular form of case 2 occurs when a is a complex number. This shows itself, for example, with an equation of the type—

$$\phi(\mathrm{E})y(x) = f(x) \cos kx$$
$$= \tfrac{1}{2}f(x)(e^{ikx} + e^{-ikx}).$$

For the present purpose we can now treat each part separately as in 2, where in the one case $a = e^{ik}$ and in the other $a = e^{-ik}$.

Examples

1. $$y(x + 2) + y(x) = \sin \tfrac{1}{2}x$$
$$(\mathrm{E}^2 + 1)y(x) = \sin \tfrac{1}{2}x.$$

The complementary function of this equation is

$$y(x) = A \, . \, (i)^x + B(-i)^x,$$

where A and B are arbitrary unit periodics. For the particular solution we need

$$y(x) = [1/(\mathrm{E}^2+1)] \sin \tfrac{1}{2}x$$
$$= [\tfrac{1}{2}/(\mathrm{E}^2+1)][\mathrm{e}^{\frac{1}{2}x} - \mathrm{e}^{-\frac{1}{2}ix}],$$

giving
$$2y(x) = \mathrm{e}^{\frac{1}{2}ix}/(\mathrm{e}^{i}+1) - \mathrm{e}^{-\frac{1}{2}ix}/(\mathrm{e}^{-i}+1)$$
$$= \frac{\mathrm{e}^{\frac{1}{2}ix} - \mathrm{e}^{\frac{1}{2}-ix} + \mathrm{e}^{\frac{1}{2}ix-i} - \mathrm{e}^{i-\frac{1}{2}ix}}{(\mathrm{e}^{i}+1)(\mathrm{e}^{-i}+1)}$$
$$= 2[\sin \tfrac{1}{2}x + \sin(\tfrac{1}{2}x - 1)]/(2 + \mathrm{e}^{i} + \mathrm{e}^{-i})$$
$$= 2[2 \sin \tfrac{1}{2}(x-1) \,.\, \cos \tfrac{1}{2}]/(2 \cos 1 + 2)$$
$$= [\sin \tfrac{1}{2}(x-1)]/\cos \tfrac{1}{2}.$$

Thus the general solution of the equation is

$$y(x) = A \,.\, (\mathrm{i})^x + B(-\mathrm{i})^x + \tfrac{1}{2}[\sin \tfrac{1}{2}(x-1)]/\cos \tfrac{1}{2}.$$
$$= C \cos(x\pi/2 + D) + \tfrac{1}{2}[\sin \tfrac{1}{2}(x-1)]/\cos \tfrac{1}{2}.$$

2. $$y(x+2) - (2 \cos \tfrac{1}{2})y(x+1) + y(x) = \cos \tfrac{1}{2}x.$$

The complementary function of this equation is the general solution of the equation

$$(\mathrm{E}^2 - 2 \cos \tfrac{1}{2} \,.\, \mathrm{E} + 1)y(x) = 0,$$

i.e.
$$y(x) = A \,.\, (\cos \tfrac{1}{2} - \sin \tfrac{1}{2})^x + B \,.\, (\cos \tfrac{1}{2} + \mathrm{i} \sin \tfrac{1}{2})^x$$
$$= C \cos(\tfrac{1}{2}x + D),$$

where C and D are unit periodics. For the particular solution we require to evaluate

$$(\mathrm{E}^2 - 2\mathrm{E} \cos \tfrac{1}{2} + 1)^{-1} \cos \tfrac{1}{2}x$$
$$= \tfrac{1}{2}(\mathrm{E} - \mathrm{e}^{\mathrm{i}/2})^{-1}(\mathrm{E} - \mathrm{e}^{-\mathrm{i}/2})^{-1}(\mathrm{e}^{\mathrm{i}x/2} + \mathrm{e}^{-\mathrm{i}x/2}).$$

The formula
$$\phi(\mathrm{E})a^x = \phi(a)a^x$$

is not applicable throughout, since when $a = \mathrm{e}^{\pm\mathrm{i}/2}$ the expression $a^2 - 2 \cos \tfrac{1}{2}a + 1$ is zero, but applying the rule to the factors that do not vanish we obtain

$$\tfrac{1}{2}(\mathrm{E} - \mathrm{e}^{\mathrm{i}/2})^{-1}(\mathrm{E} - \mathrm{e}^{-\mathrm{i}/2})^{-1}(\mathrm{e}^{\mathrm{i}x/2} + \mathrm{e}^{-\mathrm{i}x/2})$$
$$= \tfrac{1}{2}(\mathrm{E} - \mathrm{e}^{\mathrm{i}/2})^{-1}(\mathrm{e}^{\mathrm{i}/2} - \mathrm{e}^{-\mathrm{i}/2})^{-1}\,\mathrm{e}^{\mathrm{i}x/2}$$
$$+ \tfrac{1}{2}(\mathrm{E} - \mathrm{e}^{-\mathrm{i}/2})^{-1}(\mathrm{e}^{-\mathrm{i}/2} - \mathrm{e}^{\mathrm{i}/2})^{-1}\,\mathrm{e}^{-\mathrm{i}x/2}$$
$$= \frac{1}{4\mathrm{i} \sin \frac{1}{2}}\{(\mathrm{E} - \mathrm{e}^{\mathrm{i}/2})^{-1}\,\mathrm{e}^{\mathrm{i}x/2} - (\mathrm{E} - \mathrm{e}^{-\mathrm{i}/2})^{-1}\,\mathrm{e}^{-\mathrm{i}x/2}\}.$$

Let $(E - e^{i/2})^{-1} e^{ix/2} = P(x)$,

then $$e^{ix/2} = (E - e^{i/2})P(x) = P(x+1) - e^{i/2} P(x).$$

Now write $$P(x) = e^{i(x-1)/2} p(x),$$

then $$e^{ix/2} = e^{ix/2} p(x+1) - e^{ix/2} p(x),$$

or $$1 = \Delta p(x).$$

Hence we may take $p(x) = x$.

Thus $$(E - e^{i/2})^{-1} e^{ix/2} = x\, e^{i(x-1)/2}$$

and writing $-i$ for i,

$$(E - e^{-i/2})^{-1} e^{-ix/2} = x\, e^{-i(x-1)/2}.$$

So the particular solution $= \dfrac{1}{4i \sin \frac{1}{2}} \{x\, e^{i(x-1)/2} - x\, e^{-i(x-1)/2}\}$

$$= \frac{x \sin (x-1)/2}{2 \sin \frac{1}{2}},$$

a result that is easily verified.

EXERCISES (IV)

1. Prove by induction that

$$(E - \alpha)^m y_n = \alpha^{n+m} \Delta^m (y_n \alpha^{-n}).$$

2. Solve—

(i) $3y_{n+1} - y_n = 0$; $\quad y_0 = 1$.
(ii) $y_{n+1} - 3y_n = n$; $\quad y_1 = 3$.
(iii) $y_{n+1} - 2y_n = 2^{n+1}$; $\quad y_1 = 1$.
(iv) $y_{n+1} - 2y_n = n^2 . 2^n$; $\quad y_1 = 0$.
(v) $y(x+2) - 4y(x) = 5 . 3^x$.
(vi) $y(x+2) - 4y(x) = 2^x$.
(vii) $y(x+2) - 4y(x+1) + 4y(x) = 2^x$.

Simultaneous Systems of Difference Equations

So far we have confined our attention to cases where only one dependent variable occurs in the equation. Cases where more than one dependent variable is present in a system of equations can easily be reduced, in a formal way at any rate, to the preceding case.

Suppose, for example, the two equations

$$f(n, x_n, x_{n+1}, \ldots, x_{n+r}, y_n, y_{n+1}, y_{n+2}) = 0 \qquad \text{(i)}$$

$$g(n, x_n, x_{n+1}, \ldots, x_{n+r}, y_n, y_{n+1}, y_{n+2}) = 0 \qquad \text{(ii)}$$

have to be reduced to a single equation in x and n. This demands the elimination of y_n, y_{n+1}, y_{n+2}. This is clearly not possible immediately but by stepping up n to $(n+1)$ we have two additional equations

$$f(n+1, x_{n+1}, \ldots, x_{n+r+1}, y_{n+1}, y_{n+2}, y_{n+3}) = 0$$

and

$$g(n+1, x_{n+1}, \ldots, x_{n+r+1}, y_{n+1}, y_{n+2}, y_{n+3}) = 0$$

from which we have to eliminate $y_n, y_{n+1}, y_{n+2}, y_{n+3}$. This is still not possible. By stepping up once more, however, we obtain six equations from which we can now, in principle at least, eliminate the five quantities, $y_n, y_{n+1}, y_{n+2}, y_{n+3}, y_{n+4}$, and this will lead to a single difference equation in x_n, that is, in one dependent variable. This final difference equation in x_n will, in this example, be of the order $(r+2)$.

Example

To eliminate y from the equations

$$2x_n x_{n+1} = y_n - y_{n+1} + 1,$$
$$x_n^2 + x_{n+1}^2 = y_n + y_{n+1} - 1,$$

we step up n to $n+1$ giving

$$2x_{n+1}x_{n+2} = y_{n+1} - y_{n+2} + 1$$
$$x_{n+1}^2 + x_{n+2}^2 = y_{n+1} + y_{n+2} - 1.$$

Now eliminating y_n, y_{n+1}, y_{n+2} from the four equations we easily find from the first two

$$2y_{n+1} = (x_n - x_{n+1})^2 + 2,$$

and from the latter two

$$2y_{n+1} = (x_{n+1} + x_{n+2})^2.$$

Equating these gives finally

$$(x_{n+2} + x_n)(x_{n+2} + 2x_{n+1} - x_n) = 2.$$

We can now examine how many arbitrary constants, or unit periodics when the independent variable is continuous, we may expect to find in the solution of a simultaneous system of this kind in general. Let

$$f(n, x_n, x_{n+1}, \ldots, x_{n+r}, y_n, y_{n+1}, \ldots, y_{n+s}) = 0 \quad \text{(i)}$$

and

$$g(n, x_n, x_{n+1}, \ldots, x_{n+r}, y_n, y_{n+1}, \ldots, y_{n+s}) = 0 \quad \text{(ii)}$$

be the simultaneous pair.

Solving these two equations algebraically for x_{n+r} and y_{n+s}, we obtain,

$$x_{n+r} = F(n, x_n, \ldots, x_{n+r-1}, y_n, \ldots, y_{n+s-1})$$

and
$$y_{n+s} = G(n, x_n, \ldots, x_{n+r-1}, y_n, \ldots, y_{n+s-1}).$$

In order to specify x_{n+r} and y_{n+s}, we require the values of x_n, $x_{n+1}, \ldots, x_{n+r-1}$, and $y_n, y_{n+1}, \ldots, y_{n+s-1}$, in all, $(r + s)$ initial values have to be given in order that the next values of x and y may be determined. Thus, in general, the maximum number of arbitrary constants involved in the complete solution is $(r + s)$.

EXERCISE (V)

Show that if the y's are eliminated from equations (i) and (ii) by stepping up the n's as in the previous paragraph, the difference equation in x that is finally obtained is of the $(r + s)$th order, and that when the solution of this final equation in x is found, y may be determined without introducing any further arbitrary constants.

The two elementary propositions which we have established in the preceding two paragraphs, may easily be generalized for any simultaneous system of difference equations in any number of variables.

Systems of Linear Difference Equations with Constant Coefficients

The method which has been applied to eliminate all the variables except one from a system of equations in the preceding paragraphs can be used at once in the case of systems of linear difference equations, and will give rise quite clearly to a *linear* difference equation in one of the variables.

When the coefficients are constant, however, the process of elimination and the reduction to a set suitable for determining the dependent variables in succession, may be systematically developed by use of operators in a manner similar to that carried through in the case of single equations. Before discussing the general case, let us examine a particular illustration.

Consider the equations

$$x_{n+1} - 3y_n - 4x_n = 1 \qquad \text{(i)}$$

$$x_{n+1} + y_{n+1} + y_n = n \qquad \text{(ii)}$$

As in the previous paragraph we solve algebraically for x_{n+1} and y_{n+1}. Thus

$$x_{n+1} = 1 + 3y_n + 4x_n,$$

and

$$y_{n+1} = n - y_n - (1 + 3y_n + 4x_n).$$

This enables the sequence x and y to be constructed provided two initial values x_1 and y_1 are given as starting points. Thus two arbitrary constants are involved in the general solution. This suggests that the system of equations (i) and (ii) should be reducible to a single difference equation of order two in one of the dependent variables, say x_n and that the other dependent variable, y_n, may be obtained thereafter without the introduction of further arbitrary constants.

To carry this into effect let us write the equations in operational form, thus—

$$(\mathrm{E} - 4)x_n - 3y_n = 1 \qquad \text{(i}'\text{)}$$

$$\mathrm{E}x_n + (\mathrm{E} + 1)y_n = n. \qquad \text{(ii}'\text{)}$$

To eliminate y_n from these equations, we operate on (i′) with $(\mathrm{E} + 1)$, on (ii′) with 3, and add; hence

$$[(\mathrm{E} + 1)(\mathrm{E} - 4) + 3\mathrm{E}]x_n = (\mathrm{E} + 1)1 + 3n$$

or

$$\begin{vmatrix} \mathrm{E} - 4 & -3 \\ \mathrm{E} & \mathrm{E} + 1 \end{vmatrix} x_n = 2 + 3n$$

This procedure is closely analogous to that involved in the solution of a system of linear algebraic equations.

Cramer's rule for the solution of a set of p algebraic linear equations in p unknowns,

$$\begin{aligned} a_{11}x_1 + a_{12}x_2 + \dots + a_{1p}x_p &= h_1 \\ a_{21}x_1 + a_{22}x_2 + \dots + a_{2p}x_p &= h_2 \\ \dots\dots\dots\dots\dots\dots\dots\dots& \\ a_{p1}x_p + a_{p2}x_2 + \dots + a_{pp}x_p &= h_p, \end{aligned}$$

states that

$$\begin{vmatrix} a_{11} & a_{12} \dots a_{1r} \dots a_{1p} \\ a_{21} & a_{22} \dots a_{2r} \dots a_{2p} \\ \cdot & \\ \cdot & \\ \cdot & \\ a_{p1} & a_{p2} \dots a_{pr} \dots a_{pp} \end{vmatrix} x_r = \begin{vmatrix} a_{11} \dots a_{1r-1}h_1 & a_{1r+1} \dots a_{1p} \\ a_{21} \dots a_{2r-1}h_2 & a_{2r+1} \dots a_{2p} \\ \cdot & \\ \cdot & \\ \cdot & \\ a_{p1} \dots a_{pr-1}h_p & a_{pr+1} \dots a_{pp} \end{vmatrix}$$

and this can be applied to the present type of operational equations provided, of course, that the order of the operation is carefully preserved in the expansion of the determinant on the right-hand side. Thus, in our example above, we have

$$\begin{vmatrix} E-4, & -3 \\ E, & E+1 \end{vmatrix} x_n = \begin{vmatrix} 1, & -3 \\ n, & E+1 \end{vmatrix}$$
$$= (E+1)1 + 3n$$
$$= 2 + 3n.$$

The number of arbitrary constants in the solution for x_n is the degree of E in the operating determinant

$$\begin{vmatrix} E-4, & -3 \\ E, & E+1 \end{vmatrix} = (E-4)(E+1) + 3E$$
$$= E^2 - 4,$$

which, in this case is two. Thus, the equation in x_n gives

$$(E^2 - 4)x_n = 2 + 3n. \qquad \text{(iii)}$$

Hence $$x_n = A \,.\, 2^n + B \,.\, (-2)^n - (4 + 3n)/3.$$

This already introduces the two arbitrary constants, so that y_n must now be found without any further arbitrariness entering. In the present case this may be done by determining y_n from (i′) so that

$$\begin{aligned} 3y_n &= (E-4)x_n - 1 \\ &= A2^{n+1} + B(-2)^{n+1} - (7+3n)/3 \\ &\qquad - 4A2^n - 4B(-2)^n + (28+12n)/3 - 1 \\ &= -2A \,.\, 2^n - 6B(-2)^n + (7+3n) - 1. \end{aligned}$$

We may assert in fact that the equations

$$(E^2 - 4)x_n = 2 + 3n$$

and $$3y_n = (E-4)x_n - 1,$$

are completely equivalent to the original pair (i′) and (ii′) in the sense that they determine x_n and y_n with the correct degree of arbitrariness and that, in addition, they are in form suitable for the immediate determination of x_n and y_n.

This example then enables us to pose the problem of the solution of a simultaneous pair, in the following form—

Given
$$\phi_1(E)x_n + \psi_1(E)y_n = f_n \tag{1}$$
$$\phi_2(E)x_n + \psi_2(E)y_n = g_n \tag{2}$$

where $\phi_1(E)$, $\phi_2(E)$, $\psi_1(E)$, $\psi_2(E)$ are polynomial expressions in the operator E, to find a pair of the form—

$$\Phi(E)y_n = F_n \tag{3}$$
$$x_n = \chi(E)y_n + G_n \tag{4}$$

which has precisely the same number of arbitrary constants in the solution and whose solution satisfies the original pair. If this can be done the first of these enables y_n to be found immediately and the second, x_n.

To effect this we note the following characteristics of the system (1) and (2)—

(*a*) The number of arbitrary constants in the complete solution of (1) and (2) is the degree of E in

$$\begin{vmatrix} \phi_1(E), & \psi_1(E) \\ \phi_2(E), & \psi_2(E) \end{vmatrix} = \phi_1(E)\psi_2(E) - \phi_2(E)\psi_1(E).$$

This follows immediately from the discussion above.

(*b*) If $\phi_1(E)$ and $\phi_2(E)$ are any two polynomial functions of E it is always possible to find polynomial functions $P(E)$ and $Q(E)$ such that

$$\phi_1(E)Q(E) - \phi(E)P(E) = 1.$$

This follows from the fact that the breakdown into partial fractions—

$$\frac{1}{\phi_1(E)\phi_2(E)} = \frac{Q(E)}{\phi_2(E)} - \frac{P(E)}{\phi_1(E)},$$

is always possible.

(*c*) With $P(E)$ and $Q(E)$ so determined, the solution of the simultaneous pair—

$$\phi_1(E)x_n + P(E)y_n = F_n,$$
$$\phi_2(E)x_n + Q(E)y_n = G_n,$$

contains no arbitrary constants. This follows from the fact that the determinant

$$\begin{vmatrix} \phi_1(\mathrm{E}), P(\mathrm{E}) \\ \phi_2(\mathrm{E}), Q(\mathrm{E}) \end{vmatrix}$$

is of zero degree in E.

(*d*) Let x_n be eliminated from (1) and (2), giving

$$[\phi_2(\mathrm{E})\psi_1(\mathrm{E}) - \phi_1(\mathrm{E})\psi_2(\mathrm{E})]y_n = h_n. \tag{5}$$

With this we associate the equation derived by operating on (1) with $P(\mathrm{E})$, on (2) with $Q(\mathrm{E})$, and subtracting, viz.

$$[\phi_1(\mathrm{E})P(\mathrm{E}) - \phi_2(\mathrm{E})Q(\mathrm{E})]x_n + [\psi_1(\mathrm{E})P(\mathrm{E}) - \psi_2(\mathrm{E})Q(\mathrm{E})]y_n = k_n$$

or

$$x_n + [\psi_1(\mathrm{E})P(\mathrm{E}) - \psi_2(\mathrm{E})Q(\mathrm{E})]y_n = k_n. \tag{6}$$

Then (5) and (6) are equivalent to (1) and (2) in the sense that all the solutions of (1) and (2) are those of (5) and (6), and all those of (5) and (6) are also solutions of (1) and (2). To establish this we need merely show that the number of arbitrary constants involved in the solution of (5) and (6) is the same as that of (1) and (2), for (5) and (6) have been derived from (1) and (2). We have to ensure that in the process we have neither gained nor lost constants.

Now the number of constants in the solution of (5) and (6) is the degree of E in the determinant

$$\begin{vmatrix} 0 & \phi_2(\mathrm{E})\psi_1(\mathrm{E}) - \phi_1\psi_2(\mathrm{E}) \\ 1 & \psi_1(\mathrm{E})P(\mathrm{E}) - \psi_2(\mathrm{E})Q(\mathrm{E}) \end{vmatrix} = \phi_2(\mathrm{E})\psi_1(\mathrm{E}) - \phi_1(\mathrm{E})\psi_2(\mathrm{E}).$$

Thus (5) and (6) are completely equivalent to (1) and (2). Moreover it is apparent that they are now in a form particularly suitable for solution. From (5) we solve for y_n with the appropriate number of constants, and thereafter (6) provides x_n immediately.

These considerations therefore enable us to lay down a systematic procedure of an elementary nature for arriving at the solution directly for a simultaneous pair in the following manner. By direct operation we derive an equation from which x_n is absent. We solve the resulting equation for y_n. By a suitable choice of operators $P(\mathrm{E})$ and $Q(\mathrm{E})$ acting on (1) and (2) we derive an equation in which the coefficient of x_n is unity. This determines x_n explicitly.

Example

Consider

$$x_{n+1} - 2x_n + y_{n+2} + y_n = 1,$$
$$x_{n+1} + x_n + y_{n+2} + 3y_{n+1} = n,$$

or, written in operational form

$$(\mathrm{E} - 2)x_n + (\mathrm{E}^2 + 1)y_n = 1,$$
$$(\mathrm{E} + 1)x_n + \mathrm{E}(\mathrm{E} + 3)y_n = n.$$

The equation in y_n above will be found by operating on the first of these with E + 1 and on the second with E − 2 and subtracting. To obtain the other of the equivalent pair we have to operate on the first with $P(\mathrm{E})$ and on the second with $Q(\mathrm{E})$ chosen so as to make

$$\begin{vmatrix} \mathrm{E} - 1 & P(\mathrm{E}) \\ \mathrm{E} - 2 & Q(\mathrm{E}) \end{vmatrix}$$

reduce to a constant. Obviously $P(\mathrm{E}) = 1$ $Q(\mathrm{E}) = 1$ suffices. Thus the second equation is obtained merely by subtracting the first two. Accordingly

$$[(\mathrm{E} + 1)(\mathrm{E}^2 + 1) - (\mathrm{E} - 2)\mathrm{E}(\mathrm{E} + 3)]y_n$$
$$= (\mathrm{E} + 1)1 - (\mathrm{E} - 2)n = n + 1,$$

and $$- 3x_n - (3\mathrm{E} - 1)y_n = 1 - n.$$

These reduce to

$$(7\mathrm{E} + 1)y_n = n + 1$$
$$x_n = \tfrac{1}{3}(n - 1) - y_{n+1} + \tfrac{1}{3}y_n,$$

from which it follows that

$$x_n = 3A(-\tfrac{1}{7})^n + (8n + 1)/64$$
$$y_n = 10A(-\tfrac{1}{7})^n/7 + n/4 - 15/32$$

where A and B are arbitrary constants.

We note that the whole of the foregoing discussion could have been dealt with in terms of x, a current continuous variable instead of n an integer. The boundary conditions would, of course, then be of a different nature.

The solution of a system of three or more equations with a corresponding number of dependent variables can be carried through by exactly the same procedure as for two such variables, step-by-step. For example, if x, y, z be the dependent variables we can replace the original three equations by three equivalent ones in which, say, z is absent from the last pair. This pair may then be treated to obtain an equivalent pair in one of which y also is now absent. This equation is now solved for x and in succession from the final equivalent set y and then z are found, the required number of constants automatically adjusting itself during the actual solution.

We illustrate these procedures by examples.

Examples

1. Solve the system

$$x_n + (E - 1)y_n + (E - 2)z_n = 0 \tag{i}$$

$$x_n + (E - 1)^2 y_n + (E - 2)^2 z_n = 0 \tag{ii}$$

$$x_n + (E - 1)^3 y_n + (E - 2)^3 z_n = 0. \tag{iii}$$

By subtracting (i) from (ii) and (ii) from (iii) we immediately derive the following equivalent set

$$x_n + (E - 1)y_n + (E - 2)z_n = 0 \tag{i}$$

$$(E - 1)(E - 2)y_n + (E - 2)(E - 3)z_n = 0 \tag{iv}$$

$$(E - 1)^2(E - 2)y_n + (E - 2)^2(E - 3)z_n = 0. \tag{v}$$

To eliminate y_n from (iv) and (v) we operate on (iv) with $E - 1$ and subtract. We have then the following scheme—

$$\begin{vmatrix} E - 1 & X \\ 1 & Y \end{vmatrix} = \text{constant}.$$

Hence choose $X = E - 2$ and $Y = 1$ as the operators for the other member of the equivalent pair. Accordingly (iv) and (v) now provide

$$(E - 2)(E - 3)z_n = 0 \tag{vi}$$

$$(E - 1)(E - 2)y_n = 0. \tag{vii}$$

and these have to be solved in association with (i). Accordingly (vi) and (vii) give

$$z_n = A \, . \, 2^n + B \, . \, 3^n,$$
$$y_n = C + D \, . \, 2^n,$$

and (i) leads to

$$x_n = -(\mathrm{E}-1)y_n - (\mathrm{E}-2)z_n$$
$$= -D \cdot 2^n - B \cdot 3^n,$$

so that there are 4 arbitrary constants in all.

Check

The determinant

$$\begin{vmatrix} 1 & \mathrm{E}-1 & \mathrm{E}-2 \\ 1 & (\mathrm{E}-1)^2 & (\mathrm{E}-2)^2 \\ 1 & (\mathrm{E}-1)^3 & (\mathrm{E}-2)^3 \end{vmatrix}$$
$$= (\mathrm{E}-1)(\mathrm{E}-2)^2(\mathrm{E}-3),$$

which is a function of E of the 4th degree. The solution should therefore contain 4 constants associated with 1^n, 2^n, and 3^n.

2. $$2x(t+1) - 4x(t) - y(t) = 2,$$
$$x(t) + 2y(t+1) - 6y(t) = 9t - 4,$$

where $x = t$, $y = 0$ for the range $0 \leqslant t < 1$.

Thus $$2(\mathrm{E}-2)x(t) - y(t) = 2,$$
$$x(t) + 2(\mathrm{E}-3)y(t) = 9t - 4.$$

Eliminating y we obtain

$$[4(\mathrm{E}-2)(\mathrm{E}-3)+1]x(t) = 2(\mathrm{E}-3) \cdot 2 + 9t - 4$$
$$= 9t - 12,$$

i.e. $$(\mathrm{E}-\tfrac{5}{2})^2 x(t) = \tfrac{9}{4}t - 3.$$

Hence by the usual procedure we find

$$x(t) = (A + Bt)(\tfrac{5}{2})^t + t,$$

where A and B are unit periodics in t. From the first equation therefore we have

$$y(t) = 2(\mathrm{E}-2)x(t) - 2 = (5B + Bt + A)(\tfrac{5}{2})^t - 2t.$$

To determine the unit periodics A and B, we have, for the range $0 \leqslant t < 1$,

$$t = (A + Bt)(\tfrac{5}{2})^t + t$$

and

$$0 = (5B + A + Bt)(\tfrac{5}{2})^t - 2t.$$

Thus

$$A + [t]B = 0,$$

and

$$A + (5 + [t])B = 2[t](\tfrac{5}{2})^{-[t]}.$$

Accordingly

$$B = [t](\tfrac{2}{5})^{1+[t]},$$

and

$$A = -[t]^2(\tfrac{2}{5})^{1+[t]}.$$

Hence

$$x(t) = [t](\tfrac{2}{5})^{1+[t]-t}(t - [t]) + t,$$

and

$$y(t) = [t](\tfrac{2}{5})^{1+[t]-t}(t - [t] + 5) - 2t,$$

for all positive values of t.

CHAPTER 5

The General Difference Equation of the First Order

IN this chapter we propose to examine the nature of the solutions of the difference equation—

$$y(x + 1) = F[y(x)]. \qquad (1)$$

in which x does not appear explicitly, although much of what we have to say will be equally applicable to the more general form—

$$G[y(x + 1),\ y(x)] = 0.$$

It has already been pointed out that if $y(x)$ is initially prescribed for all values of x within a range of variation of unity, then it is possible in principle to construct the value of $y(x)$ at any other value of x, directly from equation (1) itself. To each value $y(x)$, there is a unique sequence $y(x + 1)$, $y(x + 2)$, $y(x + 3)$ . . . to be obtained in this way. The process can best be described graphically in the following way. We take two rectangular axes $y(x)$ horizontally and $y(x + 1)$ vertically in the usual manner, where $y(x)$ and $y(x + 1)$ extend along these axes from $-\infty$ to $+\infty$. This graph may be drawn from equation (1) and is quite determinate. For purposes of illustration let it be represented by the branch AB in Fig. 5.1. In this diagram CD is the straight line $y(x + 1) = y(x)$. Let OP_0 be the initial value $y(x)$, then P_0Q_0 is $y(x + 1)$. We draw Q_0R_0 to meet $y(x + 1) = y(x)$ in R_0, and drop a perpendicular on to the axis of $y(x)$ from R_0 meeting the curve in Q_1 and the axis in P_1. Then OP_1 which equals P_0Q_0 is $y(x + 1)$. By proceeding this way we obtain the set of points on the $y(x)$ axis, corresponding to $y(x)$, $y(x + 1)$, $y(x + 2)$, $y(x + 3)$ . . ., namely P_0, P_1, P_2, P_3. . . . We repeat that this succession of points is explicitly independent of x, depending only on the initial value OP_0 and on the curve AB. The point P_0 is a typical member of the set of points representing the initial values of $y(x)$ over a range of unit variation of x.

We notice that if the branch AB lies everywhere below CD and stretches from (∞, ∞) in the first quadrant or from $(\infty, -\infty)$ in the fourth quadrant to $(-\infty, -\infty)$ in the third quadrant, the succession of points $P_0, P_1, P_2 \ldots$ move to the left, while if AB lies above CD they move to the right. Thus if the graph of (1) stretches in this way below CD then $y(x)$ tends to $-\infty$ as x increases indefinitely, whereas if it lies above CD then $y(x)$ tends to $+\infty$, whatever be the initial value of $y(x)$.

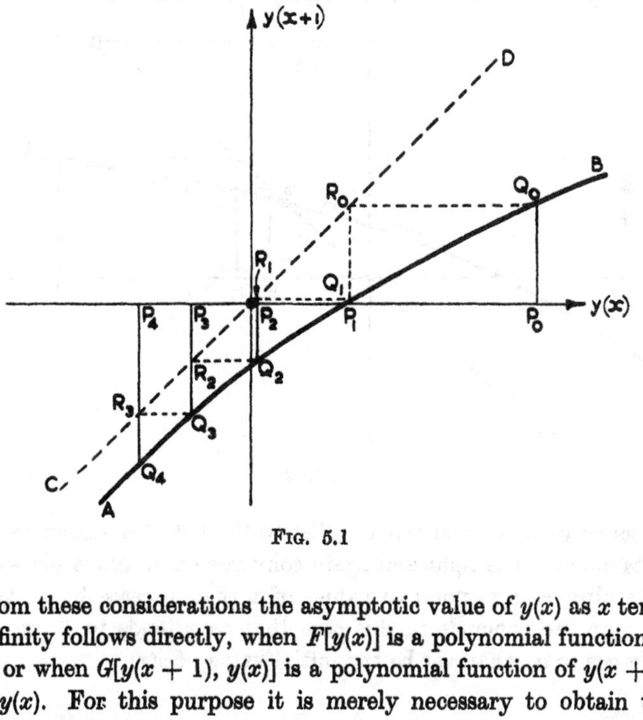

FIG. 5.1

From these considerations the asymptotic value of $y(x)$ as x tends to infinity follows directly, when $F[y(x)]$ is a polynomial function of $y(x)$, or when $G[y(x+1), y(x)]$ is a polynomial function of $y(x+1)$ and $y(x)$. For this purpose it is merely necessary to obtain the approximation to the curve at $(-\infty, -\infty)$ or at (∞, ∞) as the case may be, and solve the corresponding difference equation.

Example

$$y(x+1) = [y(x) + 2][y(x) + 1].$$

The graph in the $y(x+1)$, $y(x)$ plane is a parabola lying entirely above the line $y(x+1) = y(x)$, and the approximation at (∞, ∞) is $y(x+1) = y(x)^2$, the solution of which is $y(x) = C^{2^x}$ where C is a unit periodic. Since $y(x)$ tends to infinity as x increases indefinitely, C must be greater than 1.

We turn now to the case in which the branch crosses the line CD once, say at the point (a, a). To illustrate our point we will consider two cases—

1. AB lies above CD for all values of $y(x)$ less than a, and crosses the line CD with a positive slope less than unity. It is then clear from the diagram, (Fig. 5.2), that when P_0, the starting point, lies to the right of a, the successive points P_1, P_2 etc. move to the left

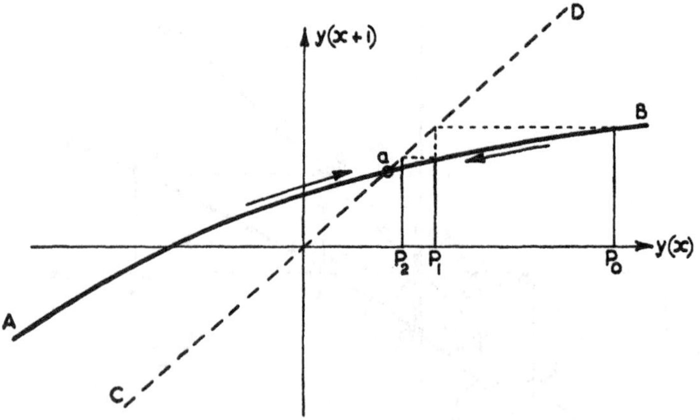

FIG. 5.2

and converge to a; and when P_0 lies to the left of a, the successive points move to the right and again converge on a. Since successive points always correspond to values of x that increase by unity at each step, we conclude in this case that as x tends to $+\infty$, $y(x)$ converges to a, whatever be the initial value of $y(x)$.

Example

$$2y(x+1) = y(x) + 1.$$

Here the limiting value is 1, and this is brought out by the standard solution $y(x) = 1 + A/2^x$.

2. AB lies below CD for all values of $y(x)$ less than a, and crosses the line CD at a with a slope greater than unity. It is then obvious by tracing out the successive positions, (Fig. 5.3), that when the starting point P_0 lies to the left of a, the successive points lie still

farther to the left without limit, and when the starting point lies to the right of a, the successive points lie still farther to the right. Thus $y(x)$ becomes infinitely large positively or negatively according as the initial boundary value lies to the right or the left of a. If P_0 coincides with a, then all succeeding positions are at a. Thus $y(x) = a$ is a special solution.

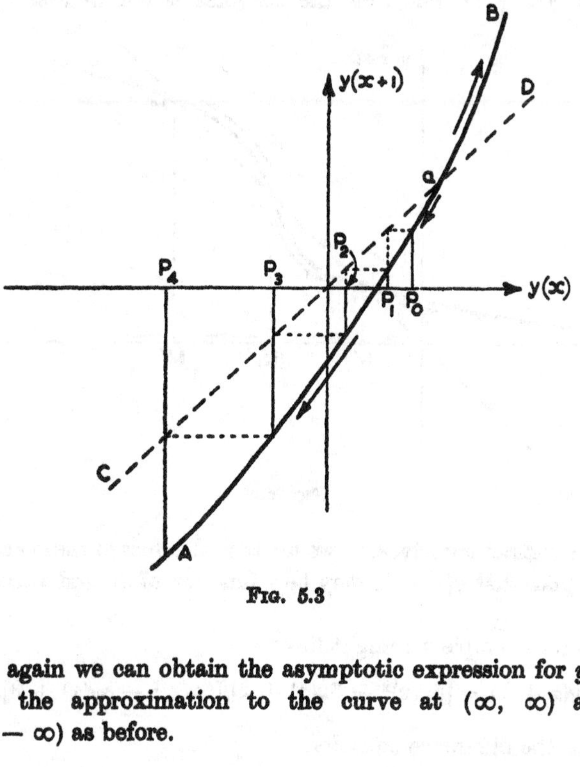

FIG. 5.3

Once again we can obtain the asymptotic expression for $y(x)$ by seeking the approximation to the curve at (∞, ∞) and at $(-\infty, -\infty)$ as before.

Example

$$y(x+1) = y(x)^3 - 6.$$

This cuts the line CD at (2, 2). Thus $y(x) = 2$ is a special solution. The approximation at (∞, ∞) is evidently

$$y(x+1) = y(x)^3$$

and the solution of this equation is c^{3^x}, where c is a unit periodic, and this is the asymptotic expression for $y(x)$ when x is large.

Appropriate Modes of Expansion

We are now in a position to examine the problem in slightly greater generality than we have done above. We assume that equation (1) defines $y(x+1)$ uniquely in terms of $y(x)$, that is that $y(x+1)$ is a single-valued function of $y(x)$, and that corresponding to every value of $y(x)$ ranging from $+\infty$ to $-\infty$ there is a definite value of $y(x+1)$, but that the converse is not necessarily true,

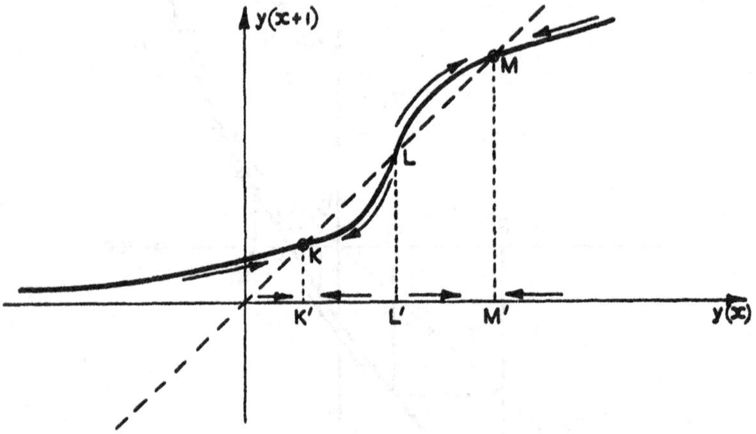

FIG. 5.4

when we restrict ourselves, as we do, to real values of these variables. This implies that $y(x+1)$ may be a function of limited variation in $y(x)$.

Thus for example we might have

$$y(x+1) = [5y(x)^2 - 6y(x) + 2]/[6y(x)^2 - 8y(x) + 3].$$

Consider the difference equation

$$y(x+1) = F[y(x)]. \tag{1}$$

with its corresponding curve in the $y(x)$, $y(x+1)$ plane as before, and let us suppose it crosses the line $y(x+1) = y(x)$ at the points $K, L, M, N, \ldots$. These points correspond to all the real solutions of the equation

$$z = F(z).$$

In Fig. 5.4 the slopes at $K, L, M, \ldots$ are assumed to be alternately less than unity and greater than unity, while in Fig. 5.5 they are assumed to be alternately greater than and less than unity, all slopes supposed for the moment to be positive.* In Fig. 5.4 initial values of $y(x)$ that lie between K' and L' will lead to a sequence of values that converge finally to K', and those that lie between L' and M' will converge to M' as indicated by the arrows. The points of convergence will correspond to alternate points K, M, O . . . on the

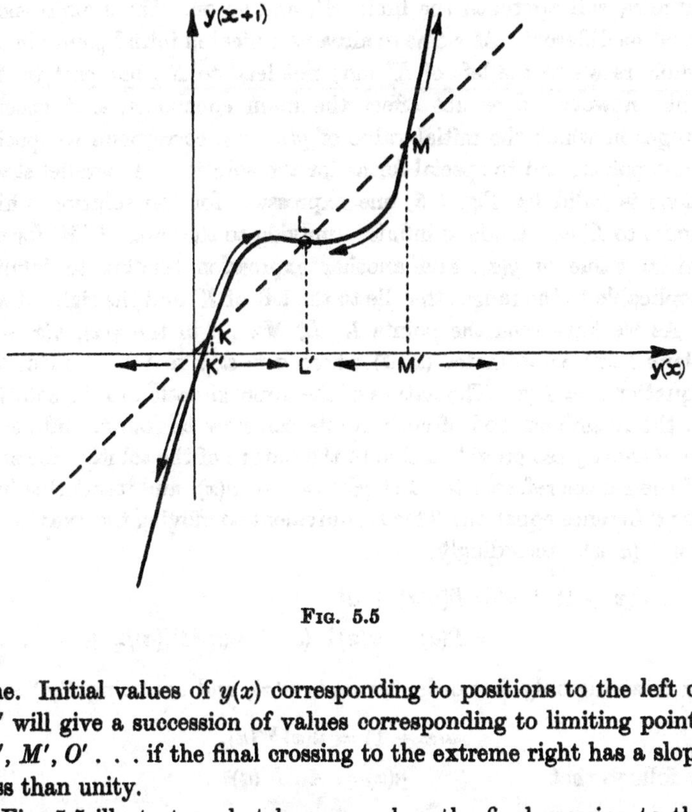

FIG. 5.5

line. Initial values of $y(x)$ corresponding to positions to the left of K' will give a succession of values corresponding to limiting points K', M', O' . . . if the final crossing to the extreme right has a slope less than unity.

Fig. 5.5 illustrates what happens when the final crossing to the right occurs with a slope greater than unity, as it also illustrates what happens when the first crossing to the left, K, occurs with a slope greater than unity. In all cases the arrows indicate the direction of x increasing without limit.

* When these assumptions are not fulfilled special treatment is required but the mode of approach here outlined is still valid and applicable once the points of convergence have been determined.

These figures bring out how closely the formal expression for the solution of the difference equation is bound up with the initial value. For example from Fig. 5.4 we expect in this case that when the given value of $y(x)$ lies to the left of L' the solution to the difference equation will be an expression in x which tends to the limit K' as x tends to infinity. When the initial value of $y(x)$ lies in the range L' to N' the solution will be an expression in x which, as x tends to infinity, will approach the limit M', and so on. These expressions must be different. As we have already indicated initial points in the whole range to the left of K' may not lead to K', but part to M'. This, however, does not affect the main conclusion that specific ranges in which the initial value of $y(x)$ lies, correspond to specific limit points, and to special forms for the solution. A parallel statement is valid for Fig. 5.5, one expression for the solution which limits to L' as x tends to infinity applying to the range $K'M'$ for the initial value of $y(x)$, and another expression tending to infinity applicable to the ranges that lie to the left of K' and the right of M'.

As we have seen the points K, L, M . . . in the $y(x)$, $y(x+1)$ plane have co-ordinates (a, a) where a is any real solution of the equation $z = F(z)$. The nature of the approximation to the solution in the neighbourhood of such points can now be found, and as we shall shortly see provide a clue to the nature of the solution for each of the ranges referred to. Let $y(x) = a + u(x)$, and insert this into the difference equation. This is equivalent to moving the axes to the point (a, a). Accordingly,

$$\begin{aligned} u(x+1) + a &= F[u(x) + a] \\ &= F(a) + u(x)F'(a) + u(x)^2F''(a)/2 + \ldots, \end{aligned}$$

or retaining only first order terms in $u(x)$, and assuming $F(a) = a$,

$$u(x+1) = u(x)F'(a).$$

It follows that

$$u(x) = A \,.\, F'(a)^x$$

where A is a unit periodic. Thus the first approximation to $y(x)$ at the limit point (a, a), where x tends to infinity, is

$$y(x) = a + AF'(a)^x,$$

where $|F'(a)| < 1$. These must correspond to the first two terms in the expression for $y(x)$ applicable to the range that contains the limit point (a, a).

A special form arises when at such a point $F'(a) = 0$, that is to say the curve in the $y(x)$, $y(x + 1)$ plane crosses the line

$$y(x + 1) = y(x)$$

horizontally. For in such circumstances

$$u(x + 1) = u(x)^2 F''(a)/2,$$

and this leads immediately to—

$$\log u(x) = A \, . \, 2^x - \log (F''(a)/2),$$

or

$$u(x) = 2c^{2^x}/F''(a).$$

Thus the asymptotic value of $y(x)$ at the limit point (a, a), as x tends to infinity, is

$$y(x) = a + 2c^{2^x}/F''(a)$$

if $F'(a) = 0$, and $a = F(a)$. Again these must correspond, in this case, to the first two terms in the expression for $y(x)$ in the range that possesses the limit point (a, a). For this purpose $|c|$ must be less than unity.

It remains finally to consider the situation envisaged in Fig. 5.5 to the left of K' and to the right of M' where the branches of the $y(x), y(x + 1)$ curve run off to $(-\infty, -\infty)$ and (∞, ∞) respectively, as x increases without limit. By use of Newton's diagram, or by any other orthodox method of approximation, the asymptotic relationship between $y(x + 1)$ and $y(x)$ is easily found, whether the original equation is in the form (1) or (2). Let this asymptotic relationship be

$$y(x + 1) = py(x)^r,$$

where r is necessarily greater than 1 since the curve is assumed ultimately to lie above $y(x + 1) = y(x)$, as $y(x)$ tends to ∞. The special case where $r = 1$ and $p < 1$ will be considered later. As before then the solution of this difference equation is easily found to be

$$y(x) = c^{r^x}/q,$$

where $q = p^{1/(r-1)}$, c is again a unit periodic, and $|c| > 1$. This must therefore represent the first term in the expansion of $y(x)$ for the range to the right of M in Fig. 5.5. The various special forms adopted by these asymptotic values are illustrated in the examples that follow.

We turn now to an examination of the actual forms of expansions applicable to the various ranges indicated for the initial values and which lead to the above determined asymptotic values. We begin by recapitulating the successive steps and the points that are significant for our present purpose.

1. The difference equation being

$$y(x+1) = F[y(x)],$$

we first sketch the curve in the $y(x)$, $y(x+1)$ plane, and by solving the equation $a = F(a)$ for real values of a, indicate the crossing points with the line $y(x+1) = y(x)$.

2. We then insert in the diagram, arrows to indicate which initial ranges correspond to which limiting points, or to branches which move off to infinity. This must be carefully done because it is not always the case that contiguous ranges of the same branch that passes to infinity have, as we have explained, the same limiting points.

3. For the purposes of 2 above, the values of $F'(a)$, and where $F'(a) = 0$, the value of $F''(a)$, have to be calculated.

4. We write down the asymptotic approximations to $y(x)$ corresponding to each of the limit points, and to the branch at infinity.

Surveying these results we see that the asymptotic approximations adopt three distinctive forms, viz.

$$y(x) = a + AF'(a)^x + \ldots$$

where $|F'(a)| < 1$, valid at (a, a).

$$y(x) = a + Bc^{2^x} + \ldots$$

when $F'(a) = 0$ where $|c| < 1$, valid at (a, a).

$$y(x) = Cc^{r^x} + \ldots$$

where $|c| > 1$, valid at (∞, ∞), with $r > 1$.

The third form differs from the second in that r may be a fraction. The second form could be

$$y(x) = a + Bc^{3^x}$$

if, for example, $F''(a) = 0$. The distinction, however, is in general immaterial to us since in any specific case the actual values of a and r are determined for us from the original equation, as we have shown. We consider each of these three cases separately.

1. $$y(x) = a + AF'(a)^x + \ldots.$$

Consider the transformation $t = AF'(a)^x$. Then

$$F'(a) \cdot t = AF'(a)^{x+1}.$$

This indicates that instead of the original equation

$$y(x+1) = F[y(x)],$$

we can take the alternative equation

$$z[F'(a)t] = F[z(t)],$$

or

$$z(kt) = F[z(t)],$$

where

$$k = F'(a),$$

and if an expression can be found for $z(t)$, then the corresponding solution to the original equation will be obtained by replacing t by $AF'(a)^x$, where a is an arbitrary unit periodic. Moreover, since the asymptotic value of $y(x)$ is $a + AF'(a)^x$, the type of solution we seek for $z(t)$ is

$$z(t) = a + t + A_2t^2 + A_3t^3 + A_4t^4 + \ldots.$$

This will satisfy all the requirements provided the coefficients A_2, A_3, A_4 etc. can be uniquely determined. We shall see shortly that under the conditions here specified this is always the case. The solution $y(x)$ will then occur in the form—

$$y(x) = a + Ak^x + A_2A^2k^{2x} + A_3A^3k^{3x} + \ldots,$$

where $k = F'(a)$ and $|k| < 1$.

2. $$y(x) = a + Bc^{2^x} + \ldots.$$

Consider the transformation $t = c^{2^x}$. Then

$$t^2 = c^{2 \cdot 2^x} = c^{2^{x+1}}.$$

This indicates that, instead of the original equation

$$y(x+1) = F[y(x)],$$

we can take the alternative equation

$$z(t^2) = F[z(t)],$$

and if an expression for $z(t)$ can be found to satisfy this equation, then the corresponding solution to the original equation is obtained by replacing t by c^{2^x}, where c is an arbitrary unit periodic. Now the asymptotic expression for $y(x)$ is

$$y(x) = a + B_1 c^{2^x} + \ldots$$

where B_1 is a definite number, $2/F''(a)$, and c is an arbitrary unit periodic such that $|c| < 1$. Thus the type of solution to be sought is

$$z(t) = a + B_1 t + B_2 t^2 + B_3 t^3 + \ldots,$$

and this will satisfy all the requirements provided the coefficients of the successive powers of t can be found uniquely. This, as we shall see shortly, is always the case under the conditions here specified. The corresponding solution will then take the form—

$$y(x) = a + B_1 c^{2^x} + B_2 c^{2 \cdot 2^x} + B_3 c^{3 \cdot 2^x} + \ldots,$$

where c is an arbitrary unit periodic and $|c| < 1$. It is to be remembered that in this case $F'(a) = 0$. If, in addition, $F''(a) = 0$ then the corresponding form would have been

$$y(x) = a + B_1 c^{3^x} + B_2 c^{2 \cdot 3^x} + B_3 c^{3 \cdot 3^x} + \ldots.$$

This would be the case when the $y(x)$, $y(x + 1)$ curve crosses the line $y(x + 1) = y(x)$ at (a, a) horizontally, and has a point of inflection at that position.

3. Here $y(x) = bc^{r^x} + \ldots$ where $r > 1$, $|c| > 1$ and b is known.

Write $t = c^{r^x}$, so that $t^r = c^{r^{x+1}}$. As before instead of the original equation we may now write

$$z(t^r) = F[z(t)],$$

and seek a solution of the form—

$$z(t) = bt + A_0 + A_1/t + A_2/t^2 + A_3/t^3 + \ldots.$$

and when the coefficients have been determined the solution of the original equation follows by replacing t by its equivalent expression in x, viz. c^{r^x}.

Now it will be recalled that this has been derived from the asymptotic relationship at (∞, ∞) in the $y(x)$, $y(x + 1)$ plane in the form—

$$y(x + 1) = py(x)^r.$$

The special case, previously referred to in this connection, arises when $r = 1$ and $p > 1$, in which case the $y(x)$, $y(x + 1)$ curve still lies above the $y(x+1) = y(x)$ line at (∞, ∞). Here, then, the asymptotic value for $y(x)$ becomes Ap^x where A is an arbitrary unit periodic. If, as before, we write $t = Ap^x$ the difference equation now takes the form—

$$z(pt) = F[z(t)],$$

and the corresponding expansion for $z(t)$ in terms of t becomes—

$$z(t) = t + A_0 + A_1/t + A_2/t^2 + A_3/t^3 + \ldots,$$

similar to the ordinary form of expansion for this section except that t has a different significance.

We propose now to show that these expansions can be derived, term by term, in a unique way, from the functional equations that now replace the original difference equation. For this purpose we restrict ourselves to the functional equations of sections 1 and 2 above in treating the general case, and deal with the functional equation of section 3 by illustration in particular cases.

By a direct application of Taylor's theorem it can be shown that if

$$f(a_0 + a_1 t + a_2 t^2 + \ldots) = A_0 + A_1 t + A_2 t^2 + \ldots,$$

then

$$A_0 = f(a_0),$$
$$A_1 = a_1 f'(a_0),$$
$$A_2 = [a_1^2 f''(a_0) + 2a_2 f'(a_0)]/2!,$$
$$A_3 = [a_1^3 f'''(a_0) + 6a_1 a_2 f''(a_0) + 6a_3 f'(a_0)]/3!,$$
$$A_4 = [a_1^4 f^{\text{iv}}(a_0) + 12a_1^2 a_2 f'''(a_0) + (24a_1 a_3 + 12a_2^2) f''(a_0) + 24a_4 f'(a_0)]/4!,$$
$$A_5 = [a_1^5 f^{\text{v}}(a_0) + 20a_1^3 a_2 f^{\text{iv}}(a_0) + 60a_1(a_1 a_3 + a_2^2) f'''(a_0) + 120(a_1 a_4 + a_2 a_3) f''(a_0) + 120 a_5 f'(a_0)]/5!.$$

1. The functional equation whose solution is required in the form of an expansion is

$$z(kt) = F[z(t)],$$

where $k = F'(a)$ and $a = F(a)$, and $z(t)$ is sought in the form

$$z(t) = a + t + a_2 t^2 + a_3 t^3 + a_4 t^4 + \ldots.$$

Inserting this expression for $z(t)$ into the functional equation we have the following requirement—

$$a + kt + a_2k^2t^2 + a_3k^3t^3 + a_4k^4t^4 + \ldots$$
$$= F(a + t + a_2t^2 + a_3t^3 + a_4t^4 + \ldots).$$

Making use of the expansion we have just set out as an extension of Taylor's theorem, and equating like powers of t on both sides we easily derive the following system of equations—

$$a = F(a), \qquad k = F'(a),$$
$$a_2k^2 = [F''(a) + 2a_2F'(a)]/2$$
$$a_3k^3 = [F'''(a) + 6a_2F''(a) + 6a_3F'(a)]/6$$
$$a_4k^4 = [F^{\text{iv}}(a) + 12a_2F'''(a) + (24a_3 + 12a_2^2)F''(a) + 24a_4F'(a)]/24.$$

It is obvious from these equations that $a_2, a_3, a_4, \ldots$ are uniquely determined as follows—

$$a_2 = F''(a)/2(k^2 - k),$$
$$a_3 = [(k^2 - k)F'''(a) + 3F''(a)^2]/6(k^3 - k)(k^2 - k)$$
$$= F'''(a)/6(k^3 - k) + F''(a)^2/2(k^3 - k)(k^2 - k),$$
$$a_4 = F^{\text{iv}}(a)/24(k^4 - k) + (3k + 5)F''(a)F'''(a)/12(k^4 - k)(k^3 - k)$$
$$+ (k + 5)F''(a)^3/8(k^4 - k)(k^3 - k)(k^2 - k).$$

With the values of $a_2, a_3, a_4, \ldots$ determined in succession in this way, it is now possible to state that given the difference equation

$$y(x + 1) = F[y(x)],$$

then if a real point (a, a) exists such that $a = F(a)$ and

$$|k| = |F'(a)| < 1,$$

then in the neighbourhood of (a, a) there exists a solution of the difference equation

$$y(x) = a + Ak^x + a_2A^2k^{2x} + a_3A^3k^{3x} + \ldots$$

where A is an arbitrary unit periodic. Although the general term in the series has not been determined it is not difficult to see from the mode of formation of the coefficients that a_n increases approximately as $1/k^{n-1}$, so that the ratio of the $(n + 1)$th term to the nth term is of the order Ak^{x-1}; and since k is less than unity in absolute value this can be made to be less than 1 by choosing x sufficiently large. The rapidity of convergence of such series will become apparent shortly when a number of examples are worked out in detail.

2. The functional equation whose solution is required in the form of an expansion is

$$z(t^2) = F[z(t)],$$

where a real value a exists such that $a = F(a)$ and $F'(a) = 0$. The form of expansion suggested is—

$$z(t) = a + b_1 t + b_2 t^2 + b_3 t^3 + \dots$$

where $b_1 = 2/F''(a)$.
Inserting this expression for $z(t)$ into the functional equation we have

$$\begin{aligned} a + b_1 t^2 + b_2 t^4 + b_3 t^6 + \dots &= F(a + b_1 t + b_2 t^2 + b_3 t^3 + \dots) \\ &= F(a) + b_1 F'(a) t + [b_1^2 F''(a) + 2b_2 F'(a)] t^2/2! \\ &\quad + [b_1^3 F'''(a) + 6b_1 b_2 F''(a) + 6b_3 F'(a)] t^3/3! \\ &\quad + [b_1^4 F^{\text{iv}}(a) + 12 b_1^2 b_2 F'''(a) + (24 b_1 b_3 + 12 b_2^2) F''(a) \\ &\qquad + 24 b_4 F'(a)] t^4/4! + \dots . \end{aligned}$$

Equating like powers on both sides gives finally—

$$\begin{aligned} a &= F(a), \\ b_1 &= 2/F''(a), \qquad F'(a) = 0, \\ b_2 &= -2F'''(a)/3F''(a)^3, \\ b_3 &= 5F'''(a)^2/9F''(a)^5 - F^{\text{iv}}(a)/3F''(a)^4 - F'''(a)/3F''(a)^3. \end{aligned}$$

As before the values of the coefficients are uniquely determined in succession in finite terms, provided $F''(a)$ is itself finite.

It is therefore now possible to state that if a real point (a, a) exists such that $a = F(a)$ and $F'(a) = 0$, then, in the neighbourhood of that point, there exists a solution to the difference equation

$$y(x + 1) = F[y(x)]$$

in the form

$$y(x) = a + b_1 c^{2^x} + b_2 c^{2 \cdot 2^x} + b_3 c^{3 \cdot 2^x} + \dots$$

where $|c|$ is a unit periodic less than 1. Such a series is rapidly convergent.

3. As has been mentioned it is not proposed to work out in detail the case of the functional equation

$$z(t^r) = F[z(t)]$$

where r is a fraction greater than unity. However, it should perhaps be mentioned that in principle there is no difference between this case and that under section 2, provided the variable t be changed appropriately. For example, if $r = 3/2$ so that the functional equation is

$$z(t^{3/2}) = F[z(t)],$$

we write $t = s^2$, and the equation takes the rational form

$$z(s^3) = F[z(s^2)],$$

the method of solution proceeding as before. For the special case $r = 1$ this step is, of course, quite irrelevant and one proceeds directly from the expansion

$$z(t) = t + A_0 + A_1/t + A_2/t^2 + \dots$$

to the step-by-step determination of the coefficients to satisfy the equation

$$z(pt) = F[z(t)].$$

Examples

1. $$y(x)y(x+1) = 2y(x)^2 + 1.$$

The graph in the $y(x)$, $y(x+1)$ plane is an hyperbola with asymptotes $y(x) = 0$ and $y(x+1) = 2y(x)$. There is no finite limiting point since $y(x+1) = y(x)$ does not meet the curve. Hence the asymptotic value is $y(x) = A2^x$. Writing $t = A2^x$ the equation becomes—

$$z(t)z(2t) = 2z(t)^2 + 1,$$

and we seek a solution

$$z(t) = t + A_0 + A_1/t + A_2/t^2 + \dots.$$

Thus

$$(t + A_0 + A_1/t + A_2/t^2 + A_3/t^3 + \dots)$$
multiplied by $(2t + A_0 + A_1/2t + A_2/4t^2 + A_3/8t^3 + \dots)$
$$= 1 + 2(t + A_0 + A_1/t + A_2/t^2 + \dots)^2.$$

Equating coefficients of equal powers of t on both sides—

$$A_0 + 2A_0 = 4A_0,$$

i.e.

$$A_0 = 0,$$

$$A_1/2 + A_0^2 + 2A_1 = 1 + 2(A_0^2 + 2A_1),$$

i.e.

$$A_1 = -2/3,$$

$$A_2/4 + A_0A_1/2 + A_0A_1 + 2A_2 = 2(2A_2 + 2A_0A_1),$$

i.e.

$$A_2 = 0,$$

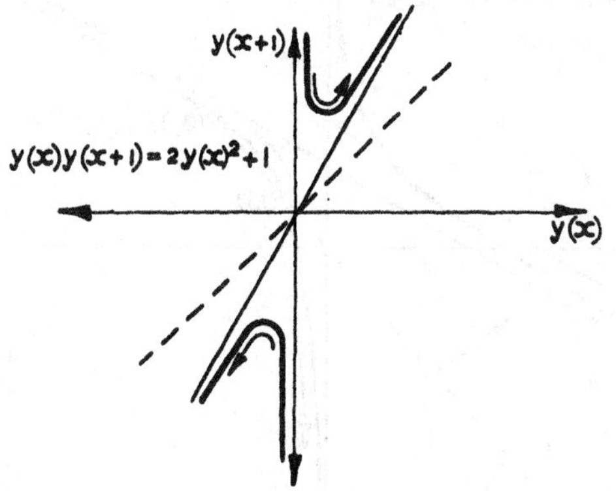

FIG. 5.6

$$A_3/8 + A_0A_2/4 + A_1^2/2 + A_0A_2 + 2A_3 = 2(A_1^2 + 2A_0A_2 + 2A_3),$$

i.e.

$$A_3 = -16/45,$$

giving

$$z(t) = t - \tfrac{2}{3}[1/t + 2^3/3\ 5t^3 + 2^6/3 . 5 . 7t^5 + \ldots].$$

Accordingly

$$y(x) = A . 2^x - \tfrac{2}{3}[2^{-x}/A + 2^{-3x} . 2^3/3 . 5A^3$$
$$+ 2^{-5x} . 2^6/3 . 5 . 7A^5 + \ldots],$$

where A is an arbitrary unit periodic (*see* Fig. 5.6).

2. $$2y(x)y(x+1) = y(x)^2 + 4y(x) - 3.$$

The graph in the $y(x)$, $y(x+1)$ plane is an hyperbola with asymptotes $y(x) = 0$ and $y(x+1) = y(x)/2 + 2$. The line $y(x+1) = y(x)$ meets the curve in (1, 1) and (3, 3) at slopes of 2 and 2/3 respectively.

All initial starting points lead finally to (3, 3). Thus the first approximation to $y(x)$ is $y(x) = 3 + A(2/3)^x$.
Accordingly write $t = A(2/3)^x$ giving a transformed equation

$$2z(t)z(2t/3) = z(t)^2 + 4z(t) - 3$$

with a possible solution of the form—

$$z(t) = 3 + t + a_2t^2 + a_3t^3 + \ldots.$$

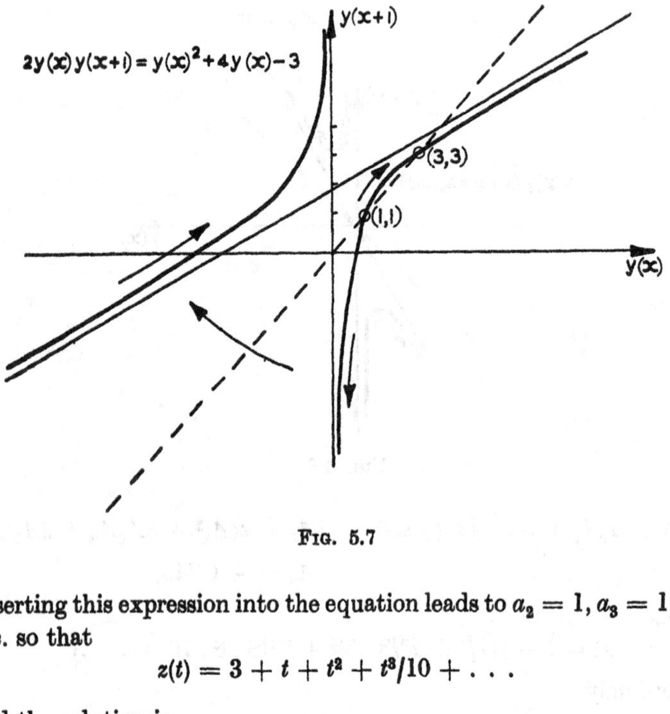

FIG. 5.7

Inserting this expression into the equation leads to $a_2 = 1$, $a_3 = 1/10$ etc. so that

$$z(t) = 3 + t + t^2 + t^3/10 + \ldots$$

and the solution is

$$y(x) = 3 + A(2/3)^x + A^2(2/3)^{2x} + A^3(2/3)^{3x}/10 + \ldots.$$

The arrows in the diagram indicate the general route followed from initial values to the limiting point (3, 3) (*see* Fig. 5.7).

3. $$y(x + 1) = y(x)^2 + a^2.$$

The curve in the $y(x)$, $y(x + 1)$ plane is a parabola. The line

$$y(x + 1) = y(x)$$

cuts the curve in the points $[1 \pm \sqrt{(1 - 4a^2)},\ 1 \pm \sqrt{(1 - 4a^2)}]$. Consider the two cases

(i) $$a^2 > \tfrac{1}{4}.$$

There are no real intersections (*see* Fig. 5.8). $y(x) \to +\infty$ indicates the general direction in which $y(x)$ proceeds from any initial value. For the asymptotic approximation we have $y(x + 1) = y(x)^2$ giving $y(x) = c^{2^x}$. Let $t = c^{2^x}$, then the equation becomes—

$$z(t^2) = z(t)^2 + a^2,$$

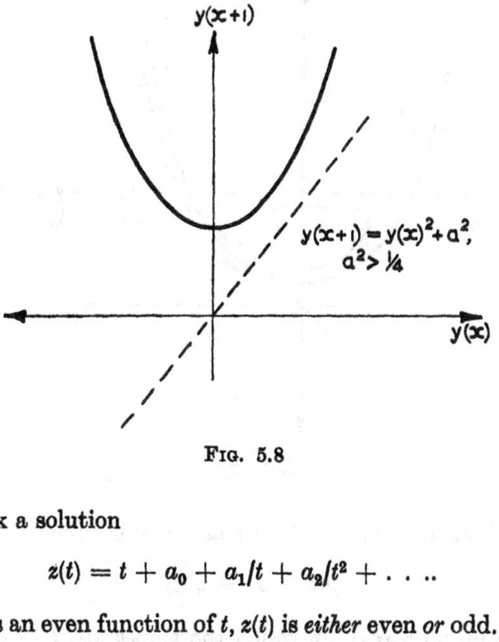

FIG. 5.8

and we seek a solution

$$z(t) = t + a_0 + a_1/t + a_2/t^2 + \ldots.$$

Since $z(t^2)$ is an even function of t, $z(t)$ is *either* even *or* odd. Since the first term in the expansion is t, it follows that

$$a_0 = a_2 = a_4 = \ldots = 0.$$

Inserting the resulting expression for $z(t)$ into the equation we easily find that—

$$a_1 = -a^2/2,$$

$$a_3 = -a^2(a^2 + 2)/8,$$

$$a_5 = -a^4(a^2 + 2)/16 \text{ etc.},$$

so that finally

$$y(x) = c^{2^x} - \frac{a^2}{2}c^{-2^x} - \frac{a^2(a^2+2)}{8}c^{-3\cdot 2^x} - \frac{a^4(a^2+2)}{16}c^{-5\cdot 2^x}$$
$$+ \frac{a^2(a^2+2)(5a^4+2a^2-8)}{128}c^{-7\cdot 2^x} + \ldots .$$

(ii) $$a^2 < \tfrac{1}{4}.$$

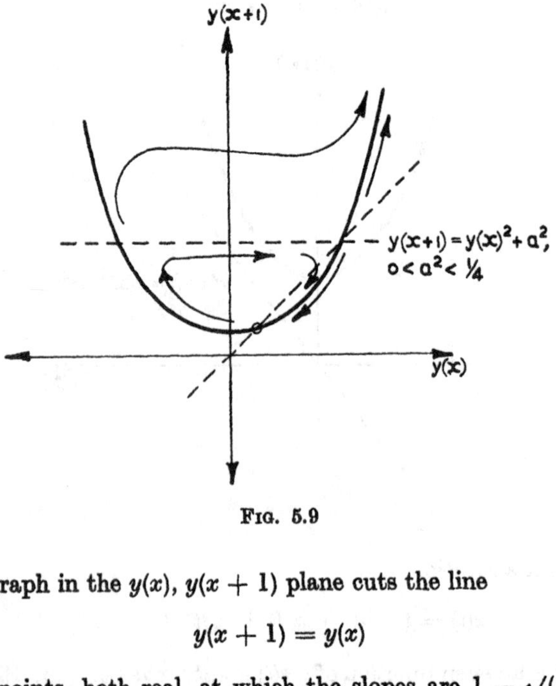

FIG. 5.9

The graph in the $y(x)$, $y(x+1)$ plane cuts the line

$$y(x+1) = y(x)$$

at two points, both real, at which the slopes are $1 - \sqrt{(1-4a^2)}$ and $1 + \sqrt{(1-4a^2)}$ in which the former is less than unity. The arrows in Fig. 5.9, indicating the direction of x increasing, show that there are two asymptotic approximations, one corresponding to the limiting point $[1 - \sqrt{(1-4a^2)}]/2$ and the other to the infinite branch related to the approximate form $y(x+1) = y(x)^2$.

Thus we have the two asympotic forms—

(a) $$y(x) = [1 - \sqrt{(1-4a^2)}]/2 + A[1 - \sqrt{(1-4a^2)}]^x$$

(b) $$y(x) = [1 + \sqrt{(1-4a^2)}]/2 + c^{2^x}.$$

In the former case we write

$$t = A[1 - \sqrt{(1 - 4a^2)}]^x$$

and the original equation takes the form—

$$z(\lambda t) = z(t)^2 + a^2,$$

where $\lambda = 1 - \sqrt{(1 - 4a^2)}$ and $0 < a^2 < \frac{1}{4}$.

In the case (*b*) above we write $t = c^{2^x}$, and the original equation becomes

$$z(t^2) = z(t)^2 + a^2.$$

The formal solution has already been found for this case in the previous section. Accordingly in

$$z(\lambda t) = z(t)^2 + a^2$$

we write $z(t) = \lambda/2 + t + a_2t^2 + a_3t^3 + \ldots,$

which we insert in the equation and equate the coefficients of equal powers of t on both sides. This leads quickly to—

$$\begin{aligned} a_2 &= -1/(\lambda - \lambda^2), \\ a_3 &= 2/(\lambda - \lambda^2)(\lambda - \lambda^3), \\ a_4 &= -(1 + 5\lambda)/(\lambda - \lambda^2)(\lambda - \lambda^3)(\lambda - \lambda^4)\ldots, \end{aligned}$$

so that the solution is

$$\begin{aligned} y(x) = \lambda/2 + A\lambda^x &- A^2\lambda^{2x}/(\lambda - \lambda^2) \\ + 2A^3\lambda^{3x}/(\lambda &- \lambda^2)(\lambda - \lambda^3) \\ &- (1 + 5\lambda)A^4\lambda^{4x}/(\lambda - \lambda^2)(\lambda - \lambda^3)(\lambda - \lambda^4) + \ldots, \end{aligned}$$

where $\lambda = 1 - \sqrt{(1 - 4a^2)} < 1.$

4. $y(x + 1) = [y(x)^2 + 3y(x) + 1]/[y(x)^2 + 4].$

The graph in the $y(x)$, $y(x + 1)$ plane has a horizontal asymptote along $y(x + 1) = 1$, cuts the line $y(x) = 0$ at $y(x + 1) = \frac{1}{4}$ and the asymptote at $(1, 1)$. This point is also the only real intersection of the graph with the line $y(x + 1) = y(x)$. The slope of the graph at $(1, 1)$ is easily found to be $3/5$. Hence the approximation is $y(x) = 1 + A(3/5)^x$. Write $t = A(3/5)^x$ then the equation to be solved is—

$$z(3t/5)[z(t)^2 + 4] = z(t)^2 + 3z(t) + 1$$

where the solution sought is

$$z(t) = 1 + t + a_2t^2 + a_3t^3 + \dots.$$

On insertion into the equation we find—

$$[1 + 3t/5 + 9a_2t^2/25 + 27a_3t^3/125 + \dots]$$

multiplied by $[5 + 2t + (1 + 2a_2)t^2 + (2a_2 + 2a_3)t^3 \dots]$

$$= [1 + 2t + (1 + 2a_2)t^2 + (2a_2 + 2a_3)t^3 + \dots] + 3[1 + t + a_2t^2 + a_3t^3 + \dots] + 1.$$

Equating coefficients of equal powers of t leads at once to $a_2 = 1$, $a_3 = 21/16 = 1{\cdot}31$, $a_4 = 737/392 = 1{\cdot}80 \dots.$

Hence

$$y(x) = 1 + A(3/5)^x + A^2(3/5)^{2x} + 1{\cdot}31A^3(3/5)^{3x} + 1{\cdot}80A^4(3/5)^{4x} + \dots.$$

5. $$y(x + 1) = y(x)[y(x)^2 + 3]/[3y(x)^2 + 1].$$

The curve in the $y(x)$, $y(x + 1)$ plane has a simple asymptote

$$y(x + 1) = y(x)/3$$

at (∞, ∞), has a point of inflection at the origin with a slope $\tan^{-1} 3$, and it crosses the line $y(x + 1) = y(x)$ at the points $(1, 1)$ and $(-1, \quad 1)$, in addition to $(0, 0)$, at zero slopes. Writing

$$y(x) = 1 + u(x),$$

and neglecting all but the lowest order terms in $u(x)$, we find

$$u(x + 1) = u(x)^3/4.$$

This leads immediately to $u(x) = 2c^{3^x}$ where c is a unit periodic. Write $t = c^{3^x}$ and the equation takes the form—

$$z(t^3)[3z(t)^2 + 1] = z(t)[z(t)^2 + 3].$$

The first approximation being

$$y(x) = 1 + u(x) = 1 + 2c^{3^x}$$

or

$$z(t) = 1 + 2t,$$

we seek a solution of the form

$$z(t) = 1 + 2t + a_2t^2 + a_3t^3 + \dots.$$

Inserting this into the equation we quickly find

$$a_2 = a_3 = a_4 = \ldots = 2,$$

where $|c| < 1$. Thus

$$z(t) = (1 + t)/(1 - t),$$

or

$$y(x) = (1 + c^{3^x})/(1 - c^{3^x}).$$

Repeating for the point $(-1, -1)$ we find that the same expression is valid there for $|c| > 1$ (*see* Fig. 5.10).

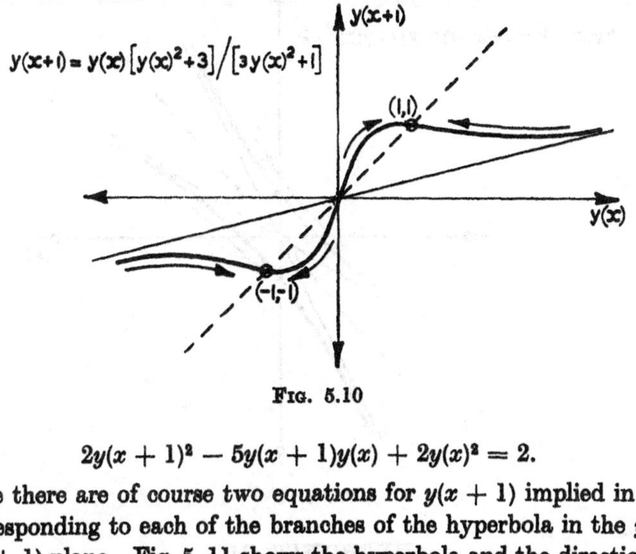

FIG. 5.10

6. $$2y(x + 1)^2 - 5y(x + 1)y(x) + 2y(x)^2 = 2.$$

Here there are of course two equations for $y(x + 1)$ implied in this corresponding to each of the branches of the hyperbola in the $y(x)$. $y(x + 1)$ plane. Fig. 5. 11 shows the hyperbola and the direction in which $y(x)$ proceeds on each from its starting point. The equation can be written

$$[2y(x + 1) - y(x)][y(x + 1) - 2y(x)] = 2,$$

so that the approximations at infinity are

(*a*) $$y(x + 1) = 2y(x) + 2/3y(x) + \ldots$$

(*b*) $$y(x + 1) = y(x)/2 - 4/3y(x) + \ldots.$$

Thus in (*a*) $y(x)$ has the asymptotic value $A2^x$, and in (*b*) $y(x)$ tends to $A/2^x$ which does not become infinite as x increases positively. For the first we write $t = A2^x$ and the equation becomes

$$[2z(2t) - z(t)][z(2t) - 2z(t)] = 2.$$

We then seek a solution of the form

$$z(t) = t + a_0 + a_1/t + a_2/t^2 + \dots.$$

Inserting this into the equation we easily find that

$$z(t) = t - 4/9t,$$

giving a solution $\qquad y(x) = A2^x - 4/9A2^x,$

where A is an arbitrary unit periodic (*see* Fig. 5.11).

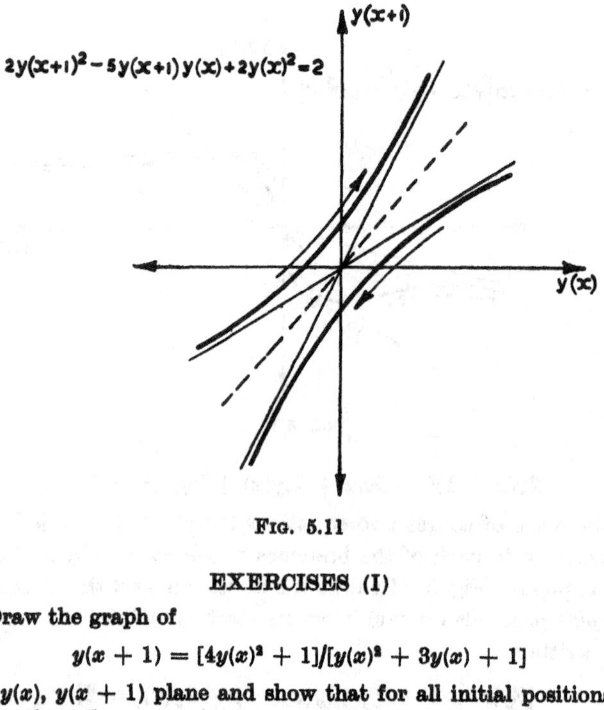

FIG. 5.11

EXERCISES (I)

1. Draw the graph of

$$y(x+1) = [4y(x)^2 + 1]/[y(x)^2 + 3y(x) + 1]$$

in the $y(x)$, $y(x+1)$ plane and show that for all initial positions, $y(x)$ limits to the value 1 as x increases indefinitely. Show further that the first three terms in the expansion of $y(x)$ are

$$1 + A(3/5)^x + 0{\cdot}3A^2(3/5)^{2x} + \dots.$$

2. Show that the solution of the equation

$$y(x+1) = y(x)^2 - 2y(x) + 2$$

is $y(x) = 1 + c^{2^x}$, and from the $y(x+1)$, $y(x)$ graph indicate the ranges of initial values of $y(x)$ which correspond to $|c| > 1$ and $|c| < 1$.

3. Determine the limiting point for the equation

$$y(x)y(x+1) = y(x)^2 + y(x) + 2.$$

Determine the first four terms in the expansion in the neighbourhood of this point, viz.

$$y(x) = -2 + A/2^x + A^2/2^{2x} + A^3/3\,.\,2^{3x} + \ldots .$$

From the $y(x)$, $y(x+1)$ diagram show that there must be another form of solution valid for positive initial values of $y(x)$.

4. Show that the difference equation

$$y(x+1) = [5y(x)^2 - 6y(x) + 2]/[6y(x)^2 - 8y(x) + 3]$$

has the two limiting points $(1, 1)$ and $(\frac{1}{2}, \frac{1}{2})$. Determine the appropriate expansions at these points and show that they both sum to the single expression

$$y(x) = (c^{2^x} + 1)/(c^{2^x} + 2)$$

where $|c| > 1$ for the region of initial values appropriate to $(1, 1)$, and $|c| < 1$ for that appropriate to $(\frac{1}{2}, \frac{1}{2})$.

5. Draw the graph of the equation

$$y(x+1) = [ay(x) + 1]/[y(x) + a]$$

in the $y(x)$, $y(x+1)$ plane; show that either $(1, 1)$ or $(-1, -1)$ is a limiting point; determine the expansion to the solution in each case; and show that both solutions can be summed in the form—

$$y(x) = [1 + A(a-1)^x/(a+1)^x]/[1 - A(a-1)^x/(a+1)^x],$$

where A is a unit periodic.

6. Show that the equation

$$5y(x+1) = [5y(x)^2 + 6y(x) + 19]/[y(x)^2 + 5]$$

has a solution of the form

$$y(x) = 1 + A/5^x + A^2/12\,.\,5^{2x-1} - A^3/78\,.\,5^{3x-2} + \ldots$$

valid for any initial value of y.

7. Obtain the first three terms in the expansion for the solution of the equation

$$8y(x+1) = [8y(x)^2 + 3]/[y(x)^2 + 1]$$

in the form

$$y(x) = \tfrac{1}{2} + A(2/5)^x + A^2(2/5)^{2x}/10 + \ldots .$$

If $y(0) = 0{\cdot}6$, find $y(3)$ and compare the result with that obtained directly from the equation.

8. From the $y(x)$, $y(x+1)$ diagram for the equation

$$y(x+1) = [4y(x)^2 + 3y(x)]/[4y(x)^2 + 3]$$

show that there is only one limiting point, and that the solution is approximately

$$y(x) = 1 + A(3/7)^x + 2A^2(3/7)^{2x} + 47A^3(3/7)^{3x}/10.$$

9. Determine two expansions for the solution of

$$y(x+1) = [y(x)^2 + ab]/[2y(x) + b - a]$$

associated with the limiting points (a, a) and $(-b, -b)$ and show that both sum to the expression

$$y(x) = (ac^{2^x} + b)/(c^{2^x} - 1).$$

10. Show that the equation

$$y(x)y(x+1) = y(x)^2 + qy(x) + q^2$$

has a solution of the form

$$y(x) = -q[1 + t + t^2/2 + 3t^3/8 + t^4/4 + 31t^5/128 + \ldots]$$

where $t = C^{2^x}$.

11. Show that the equation

$$y(x+1) = y(x) - y(x)^2$$

has the solution—

$$y(x) = -c^{2^x} + 1/2 + c^{-2^x}/8 + 9c^{-3\cdot 2^x}/128 + 9c^{-5\cdot 2^x}/1024 + \ldots.$$

If $y(0) = -2{\cdot}5$ determine $y(3)$ correct to three figures at least and compare the result with that obtained directly from the difference equation by calculating $y(1)$, $y(2)$, and $y(3)$ in succession.

12. Find the approximation—

$$y(x) = q[1 + t + t^2 + (2p + 1)t^3/8]$$

where $t = pc^{2^x}$, to the equation—

$$y(x)y(x+1) = py(x)^2 + qy(x) + pq^2/(1 - 2p)^2,$$

and determine the range of initial values of $y(x)$ for which it is valid.

13. Derive the solution

$$y(x) = \tfrac{1}{2}(x + A + 1)(x + A)$$

to the equation—

$$y(x+1) + y(x) = [y(x+1) - y(x)]^2.$$

Show that there is another solution, viz.

$$y(x) = \tfrac{1}{2}[A^2(-1)^{2x} + A(-1)^x],$$

and determine the range of initial values of $y(x)$ for which each is valid.

14. Determine in which circumstances, if any, the two functional equations—

$$y(\lambda x) = ay(x)^2 + 2hy(x) + b,$$
$$y(x^2) = ay(x)^2 + 2hy(x) + b,$$

possess solutions which can be expanded in a Taylor's Series.

15. Show that the functional equation—

$$y(x)y(\lambda x) = y(x) + 2,$$

has a solution expansible in a Taylor's Series if and only if, $\lambda = -2$ or $\lambda = -1/2$. Determine the expansions in these two cases, and show that they sum to $(2x - c_1)/(x + c_1)$ and $(2c_2 + x)/(c_2 - x)$ where c_1 and c_2 respectively satisfy the equations—

$$c_1(2x) = c_1(x),$$
$$c_2(-x/2) = c_2(x).$$

16. Derive the solution—

$$y(x) = A[\,1 + \sum_{1}^{\infty} \{2^n/x^n \prod_{1}^{n} (2^n - 1)\}] = A \prod_{0}^{\infty} (1 + 1/2^n x),$$

where $A(2x) = A(x)$, to the equation

$$(x + 1)y(2x) = xy(x).$$

17. Determine under what conditions the equation

$$y(x + 1) = [py(x)^2 + 2qy(x) + r]/[Py(x)^2 + 2Qy(x) + R]$$

may be thrown into the form—

$$[y(x + 1) + \lambda]/[y(x + 1) + \mu] = A[y(x) + \lambda]^2/[y(x) + \mu]^2.$$

Solve completely

$$y(x)^2 - 2y(x)y(x + 1) - 3y(x + 1) - 2 = 0.$$

18. In the expression $y_1(x) = x^2 - 2$ we replace x by $y_1(x)$, and write

$$y_2(x) = y_1(x)^2 - 2 = (x^2 - 2)^2 - 2 = x^4 - 4x^2 + 2.$$

This process is repeated n times, so that

$$y_{n+1}(x) = y_n(x)^2 - 2.$$

Show that $y_n(x)$ is given by the polynomial

$$2 \cos [2^{n-1} \cos^{-1} (x^2 - 2)/2].$$

19. Show that if $y_1(x) = 2\sqrt{x}$, and

$$xy_{n+1}^2(x) - (x^2 + 1)y_n(x)y_{n+1}(x) + xy_n^2(x) + (x^2 - 1)^2 = 0,$$

then

$$y_n(x) = \sqrt{x}[x^{n-1} + 1/x^{n-1}].$$

20. The function $y_n(x)$ satisfies the relation

$$4y_{n+1}(x)^2 - 10y_n(x)y_{n+1}(x) + 4y_n(x)^2 = 9.$$

If $y_1(x) = 2 \sinh x$, show that

$$y_n(x) = 2 \sinh [x + (n - 1) \log 2].$$

21. Show that the point (1, 1) in the $y(x + 1)$, $y(x)$ plane is a limiting point for the difference equation—

$$y(x + 1)^2 = y(x),$$

and that the approximation at that point is $1 + A/2^x$. Hence obtain an expansion to the solution in the form of a series of powers of $A/2^x$, and so obtain the solution $y(x) = c^{1/2^x}$.

22. Obtain

$$y(x) = 3 + A/6^x + A^2/30 \,.\, 6^{2x} + A^3/3150 \,.\, 6^{3x} + \ldots$$

from the equation

$$y(x + 1)^2 = y(x) + 6.$$

23. Show that if—

$$y_n = \sqrt{(2 + \sqrt{(2 + \sqrt{(2 + \ldots)}}}$$

where n square roots are taken, so that $y_1 = \sqrt{2}$, then

$$y_n = 2 \cos (\pi/2^{n+1}).$$

24. Find the difference equation satisfied by the continued fraction

$$y_n = 2a - b/[2a - b/2a - b/ \ldots n \text{ times}$$

where $b = a^2 - r^2$.

Show that $(a + r, a + r)$ is a limiting point in the y_{n+1}, y_n diagram, and find the expansion for y_n in the form

$$a + r + A\left(\frac{a - r}{a + r}\right)^n + a_2 A^2 \left(\frac{a - r}{a + r}\right)^{2n} + \ldots.$$

Types Soluble in Closed Form

In this section we propose to show that certain types of equation of the kind discussed in this chapter may be solved in closed form.

A. $$y(x + 1) = [ay(x) + b]/[cy(x) + 1].$$

Let $$y(x) = u(x)/v(x),$$

so that

$$u(x + 1)/v(x + 1) = [au(x) + bv(x)]/[cu(x) + v(x)].$$

Now assume that

$$u(x + 1) = au(x) + bv(x),$$

i.e. $$(\text{E} - a)u(x) - bv(x) = 0.$$

$$v(x + 1) = cu(x) + v(x),$$

i.e. $$- cu(x) + (\text{E} - 1)v(x) = 0.$$

These constitute a pair of simultaneous equations for $u(x)$ and $v(x)$. Let the roots of $(\lambda - a)(\lambda - 1) - bc = 0$ be λ_1 and λ_2, then $v(x) = A\lambda_1^x + B\lambda_2^x$ where A and B are unit periodics, and

$$u(x) = [v(x + 1) - v(x)]/c = [A(\lambda_1 - 1)\lambda_1^x + B(\lambda_2 - 1)\lambda_2^x]/c.$$

Hence $$y(x) = u(x)/v(x)$$

$$= [(\lambda_1 - 1) + K(\lambda_2/\lambda_1)^x(\lambda_2 - 1)]/c[1 + K(\lambda_2/\lambda_1)^x]$$

where K is an arbitrary unit periodic.

The direct method of this chapter may, of course, be equally applied to equations of the foregoing type. We illustrate both methods as follows.

Example $$y(x+1) = 2y(x)/[y(x)+1].$$

In the $y(x)$, $y(x+1)$ plane the line $y(x+1) = y(x)$ meets the graph in (0, 0) and (1, 1) at which the slopes are respectively 2 and $\frac{1}{2}$. Thus the first approximation to $y(x)$ is $1 + A/2^x$. Let $t = A/2^x$, then the equation becomes

$$z(\tfrac{1}{2}t)[z(t)+1] = 2z(t)$$

where we seek a solution

$$z(t) = 1 + t + A_2t^2 + A_3t^3 + \dots .$$

Inserting into the equation

$$[1 + t/2 + A_2t^2/4 + A_3t^3/8 + A_4t^4/16 + \dots]$$
$$\text{multiplied by } [2 + t + A_2t^2 + A_3t^3 + \dots]$$
$$= 2 + 2t + 2A_2t^2 + 2A_3t^3 + \dots .$$

Hence
$$A_2 + 1/2 + A_2/2 = 2A_2, \quad \text{i.e. } A_2 = 1$$
$$A_3 + A_2/2 + A_2/4 + A_3/4 = 2A_3, \quad \text{i.e. } A_3 = 1$$
$$A_4 + A_3/2 + A_2^2/4 + A_3/8 + A_4/8 = 2A_4, \quad \text{i.e. } A_4 = 1$$
$$\dots\dots\dots\dots\dots\dots\dots\dots$$

Thus $$z(t) = 1 + t + t^2 + t^3 + \dots = 1/(1-t)$$

and $$y(x) = 1/(1 - A\,.\,2^{-x}).$$

which is also the solution derived by the method of the present section.

Thus write $$y(x+1) = 2y(x)/[y(x)+1],$$

and $$y(x) = u(x)/v(x)$$

so that $$u(x+1)/v(x+1) = 2u(x)/[u(x)+v(x)].$$

Let $$u(x+1) = 2u(x),$$

i.e. $$u(x) = B\,.\,2^x$$

and $$v(x+1) = u(x) + v(x)$$

or $$v(x+1) - v(x) = B\,.\,2^x$$

so that $$v(x) = c + B\,.\,2^x.$$

Hence $$y(x) = B\,.\,2^x/[c + B2^x] = 1/[1 + c2^{-x}/B]$$
$$= 1/[1 - A2^{-x}],$$

where A is an arbitrary unit periodic.

EXERCISES (II)

1. Determine the solution of

$$y(x+1) = [1 + y(x)]/[2 - y(x)]$$

in the form $$y(x) = [1 + A2^x]/[2 - A \cdot 2^x]$$

where A is an arbitrary unit periodic.

2. If $y(x+1) = [13y(x) - 3]/[11 - 5y(x)]$, and if $y(0) = 1$, show that

$$y(6) = -67/59.$$

3. Given that $y(x+1) = [(a+3b)y(x) - 2a]/[b + 3a - 2by(x)]$ and that $y(0) = 2$, show that

$$y(3) = 2[9b - 4a]/[9a - 17b],$$

by solving the equation; and check the result directly from the equation.

B. Consider—

$$y(x+1) = [py(x) + qz(x) + r]/[Ay(x) + Bz(x) + C],$$

$$z(x+1) = [Py(x) + Qz(x) + R]/[Ay(x) + Bz(x) + C],$$

where $P, Q, R, p, q, r, A, B, C$ are constants. Write

$$y(x) = u(x)/w(x),$$

$$z(x) = v(x)/w(x),$$

then $$\frac{u(x+1)}{pu(x) + qv(x) + rw(x)} = \frac{v(x+1)}{Pu(x) + Qv(x) + Rw(x)} = \frac{w(x+1)}{Au(x) + Bv(x) + Cw(x)}.$$

Since three functions $u(x)$, $v(x)$, and $w(x)$ have been introduced in place of the unknowns $y(x)$ and $z(x)$, we may equate these ratios to unity, hence deriving the following system of linear equations—

$$(\mathrm{E} - p)u(x) - qv(x) - rw(x) = 0,$$

$$-Pu(x) + (\mathrm{E} - Q)v(x) - Rw(x) = 0,$$

$$-Au(x) - Bv(x) + (\mathrm{E} - C)w(x) = 0.$$

From these in the usual way we derive $u(x)$, $v(x)$, and $w(x)$, and thereafter their ratios to give $y(x)$ and $z(x)$ in terms of $(\lambda_1/\lambda_3)^x$ and $(\lambda_2/\lambda_3)^x$ where $\lambda_1, \lambda_2, \lambda_3$ are the roots of—

$$\begin{vmatrix} \lambda - p & -q & -r \\ -P & \lambda - Q & -R \\ -A & -B & \lambda - C \end{vmatrix} = 0.$$

The method is clearly capable of generalization to n dependent variables.

Example

$$y(x+1) = \frac{-y(x)+1}{y(x)-z(x)+1},$$

$$z(x+1) = \frac{17y(x)+6z(x)+1}{y(x)-z(x)+1}.$$

Writing $$y(x) = u(x)/w(x)$$
and $$z(x) = v(x)/w(x)$$
we have

$$\frac{u(x+1)}{-u(x)+w(x)} = \frac{v(x+1)}{17u(x)+6v(x)+w(x)} = \frac{w(x+1)}{u(x)-v(x)+w(x)}$$

Equating these to unity gives—

$$\begin{aligned}(\mathrm{E}+1)u(x) - w(x) &= 0,\\ -17u(x) + (\mathrm{E}-6)v(x) - w(x) &= 0,\\ -u(x) + v(x) + (\mathrm{E}-1)w(x) &= 0.\end{aligned}$$

Thus λ_1, λ_2, λ_3 are the roots of the cubic equation

$$\begin{vmatrix} \lambda+1 & 0 & -1 \\ -17 & \lambda-6 & -1 \\ -1 & 1 & \lambda-1 \end{vmatrix} = 0,$$

i.e. $$\lambda = -2, 3, 5.$$

Hence
$$\begin{aligned}u(x) &= A(-2)^x + B\,.\,3^x + C\,.\,5^x,\\ v(x) &= -2A(-2)^x - 7B\,.\,3^x + 2C\,.\,5^x,\\ w(x) &= -A(-2)^x + 4B\,.\,3^x + 6C\,.\,5^x,\end{aligned}$$

giving finally

$$y(x) = [1 + B_1(-3/2)^x + C_1(-5/2)^x]/ \\ [-1 + 4B_1(-3/2)^x + 6C_1(-5/2)^x],$$

$$z(x) = [-2 - 7B_1(-3/2)^x + 2C_1(-5/2)^x]/ \\ [-1 + 4B_1(-3/2)^x + 6C_1(-5/2)^x],$$

where B_1 and C_1 are unit periodics.

EXERCISES (III)

Solve—

1. $y(x+1) = [2y(x) + qz(x)]/[3y(x) - qz(x) + 5]$,
   $z(x+1) = [y(x) + 3z(x) + 1]/[3y(x) - qz(x) + 5]$.
2. $y(x+1) = [2y(x) + r]/[y(x) + z(x) + 3]$,
   $z(x+1) = [5y(x) + 7z(x) - r]/[y(x) + z(x) + 3]$.

CHAPTER 6

Linear Difference Equations with Variable Coefficients

In this and the succeeding chapter we turn to a consideration of the linear difference equation with variable coefficients. For simplicity, we shall deal only with first and second order equations, the extension to the general case will be made on each occasion as it arises and dealt with separately if it should require treatment different from that given to the equation of lower order.

The First Order Equation

The general first order linear difference equation in n may be written

$$y_{n+1} - p_n y_n = R_n \tag{1}$$

where p_n and R_n are given and defined for all integral values of n. In the first instance we consider the case where R_n is absent, so that we have the special form

$$y_{n+1} - p_n y_n = 0 \tag{1'}$$

Clearly if y_1 were specified this relation leads directly to the determination of all y's with higher suffixes. This implies, of course, that we may regard y_1 as an arbitrary constant and may expect to find it being involved in the general solution in this way. In fact we have—

$$\begin{aligned} y_n &= p_{n-1}y_{n-1}, \\ y_{n-1} &= p_{n-2}y_{n-2}, \\ y_{n-2} &= p_{n-3}y_{n-3}, \\ &\cdots\cdots\cdots \\ y_3 &= p_2 y_2, \\ y_2 &= p_1 y_1. \end{aligned}$$

Multiplying the left-hand sides of these equations together and equating the result to the product of the right-hand sides, we obtain

$$y_n = y_1 \cdot p_1 \cdot p_2 \cdot \cdot \cdot p_{n-2} \cdot p_{n-1}$$

$$= y_1 \prod_1^{n-1} p_r,$$

or $$y_n / \prod_1^{n-1} p_n = y_1,$$

where y_1, the value of y_n when $n = 1$, is so far unspecified and can therefore be regarded as an arbitrary constant. Since this expression for y_n is a function of n containing one arbitrary constant, it is in fact the general solution of the first order equation (1′).

It follows that we can make the left-hand side of (1′) a complete difference by dividing throughout by $\prod_1^n p_n$ and so transforming the equation in y_n to an equation with constant coefficients in the variable $y_n / \prod_1^{n-1} p_n$. Thus the equation

$$y_{n+1} - p_n y_n = 0,$$

on dividing by $\prod_1^n p_n$ becomes

$$(y_{n+1} / \prod_1^n p_n) - (y_n / \prod_1^{n-1} p_n) = 0,$$

$$\Delta(y_n / \prod_1^{n-1} p_n) = 0.$$

Hence $$y_n = A \,.\prod_1^{n-1} p_n,$$

where A is an arbitrary constant.

The general linear equation of the first order, viz.

$$y_{n+1} - p_n y_n = R_n \tag{1}$$

where p_n and R_n are given functions of n, can now be dealt with in exactly the same way.

Thus equation (1) leads to

$$(y_{n+1} / \prod_1^n p_n) - (y_n / \prod_1^{n-1} p_n) = R_n / \prod_1^n p_n,$$

$$\Delta(y_n / \prod_1^{n-1} p_n) = R_n / \prod_1^n p_n.$$

Hence $$y_n = A \cdot \prod_1^{n-1} p_n + (\prod_1^{n-1} p_n) \sum_1^{n-1} (R_n/\prod_1^{n} p_n),$$

which is an expression for y_n containing one arbitrary constant, and is therefore the most general solution of equation (1).

Examples

1. $$y_{n+1} - ny_n = 1.$$

Dividing through by $n!$ we obtain,

$$(y_{n+1}/n!) - (y_n/(n-1)!) = 1/n!,$$

i.e. $$\Delta(y_n/(n-1)!) = 1/n!.$$

Hence $$y_n = [\sum_1^{n-1} 1/n! + A]\,(n-1)!$$

where A is an arbitrary constant.

2. $$y_{n+1}^2 - a^n y_n^2 = a^{\frac{1}{2}n(n-1)}.$$

We write $$y_n^2 = z_n$$

thus $$z_{n+1} - a^n z_n = a^{\frac{1}{2}n(n-1)}.$$

Dividing through by $\prod_{n=1}^{n} a^n = a^{\frac{1}{2}n(n+1)}$, we obtain

$$[z_{n+1}/a^{\frac{1}{2}n(n+1)}] - [z_n/a^{\frac{1}{2}n(n-1)}] = a^{-n},$$

i.e. $$[z_n/a^{\frac{1}{2}n(n-1)}] = \sum_1^{n-1} a^{-n} + C'$$

$$= a^{-n+1}/(a-1) + C.$$

Hence $$y_n^2 = a^{\frac{1}{2}n(n-1)}[a^{-n+1}/(a-1) + C].$$

3. $$y_{n+1} - (n+1)y_n = (-1)^n \cdot n!.$$

Dividing through by $(n+1)!$ we obtain,

$$\Delta(y_n/n!) = (-1)^n/(n+1).$$

Hence, $$y_n = n!\,[\sum_1^{n-1} \{(-1)^n/(n+1)\} + C]$$

where C is an arbitrary constant.

EXERCISES (I)

Solve the equations—

1. $y_{n+1} - n^2 y_n = 0.$
2. $y_{n+1} - n y_n/(n + 1) = 0.$
3. $y_{n+1} - a^{2n+1} y_n = 0.$
4. $a^{2n+1} y_{n+1} - y_n = 0.$
5. $y_{n+1} - y_n/(n + 1) = (n - 1)/n(n + 1).$
6. $[y_{n+1} - (n + 1)]/[y_n - n] = e^{n} . n!.$
7. $n^2 y_{n+1} - (n + 1)^2 y_n = n(n + 1).$
8. $(n + 1) y_{n+1} - n y_n = n(n + 1).$
9. $y_{n+1} - e^{2n} y_n = 3n^2 e^{n^2+n}.$

As we have pointed out before, difference equations have their own special simplicities and difficulties compared with differential equations. For example, it is much easier to transform the dependent variable y_n than in the case of the differential equation, while on the other hand it is much more difficult to transform the independent variable n. As we have seen, the equation

$$y_{n+1}^2 - a^n y_n^2 = a^{\frac{1}{2}n(n-1)},$$

which is apparently non-linear, is immediately simplified by writing $z_n = y_n^2$ and the equation becomes

$$z_{n+1} - a^n z_n = a^{\frac{1}{2}n(n-1)}.$$

If on the other hand we wish to change the independent variable n to, say, $t = n^2$, the result is much more complex as we shall see. It is obvious, for example, that when n increases by unity, t does not do so and the transformed equation would not be of the form

$$F(y_t, y_{t+1}, \ldots) = 0.$$

Change of Dependent Variable

Examples

Solve—

1. $$[y_{n+2} - (n + 1)y_{n+1}] - n[y_{n+1} - n y_n] = 0.$$

Write $z_n = y_{n+1} - n y_n$, then the equation becomes

$$z_{n+1} - n z_n = 0,$$

giving $$z_n = A(n - 1)!.$$

Hence $$y_{n+1} - ny_n = A(n-1)!.$$

Dividing both sides by $n!$

$$\Delta[y_n/(n-1)!] = A/n,$$

then $$y_n/(n-1)! = A\sum_1^{n-1} 1/n + B,$$

there being two arbitrary constants since the original equation is of the second order.

2. $$y_{n+1}/y_n = e y_{n+1}^{-1/n}.$$

Taking logarithms of both sides

$$\log y_{n+1} - \log y_n = 1 - \frac{1}{n}\log y_{n+1}$$

Let $$z_n = \log y_n,$$

then $$z_{n+1} - z_n = 1 - \frac{1}{n} z_{n+1},$$

or $$(n+1)z_{n+1} - nz_n = n,$$

or $$\Delta(nz_n) = n.$$

Thus $$nz_n = \sum_1^{n-1} n + A = \tfrac{1}{2}n(n-1) + A,$$

and $$y_n = e^{\frac{1}{2}(n-1)+A/n} = e^{\frac{1}{2}(n-1)}B^{1/n}$$

where B is an arbitrary constant.

3. $$y_{n+1} = 2y_n\sqrt{(1-y_n^2)}.$$

Let $$y_n = \sin\theta_n,$$

then $$y_{n+1} = \sin\theta_{n+1}.$$

Hence $$\sin\theta_{n+1} = 2\sin\theta_n\cos\theta_n = \sin 2\theta_n$$

$$\theta_{n+1} = (-1)^r 2\theta_n + r\pi,$$

where r is any integer. When $r = 2s$, an even integer,

$$\theta_{n+1} - 2\theta_n = 2s\pi,$$

so that $$\theta_n = A2^n - 2s\pi,$$

$$y_n = \sin(A2^n - 2s\pi) = \sin(A \,.\, 2^n).$$

When $r = 2s+1$ an odd integer,

$$\theta_{n+1} + 2\theta_n = (2s+1)\pi.$$

Then $$\theta_n = A(-2)^n + (2s+1)\frac{\pi}{3},$$

and $$y_n = \sin\left[A(-2)^n + (2s+1)\frac{\pi}{3}\right].$$

The case of the continuous independent variable x instead of the discrete variable n requires special examination. This arises in the first place from the fact that the solution can be expressed, in the case of n, as a summation which can be evaluated in closed form; no such procedure can be applied in the case of x. There is no fixed lower limit to the summation which when increased by unit amounts will always lead to $x - 1$ as the upper limit to the summation, since x may have any value and not merely integral values.

Thus if we have the equation

$$y(x+1) - y(x) = x,$$

we cannot immediately write the solution in the form analogous to the case of n, viz.

$$y(x) = x(x+1)/2 + A$$

even when A is a unit periodic.

Nevertheless the two cases are more alike, in fact, than it might seem at first sight. In this case, for example, we might follow the procedure previously adopted as far as is permissible in this way—

$$y(x+1) - y(x) = x = N + [x],$$

where N is the integral part of x and $[x]$ is the fractional part. Then,

$$y(x) - y(x-1) = x - 1 = N - 1 + [x],$$

$$\cdots\cdots\cdots\cdots\cdots\cdots\cdots\cdots$$

$$y(1+[x]) - y([x]) = [x] = 0 + [x].$$

Then, summing, we obtain

$$\begin{aligned} y(x+1) - y([x]) &= N(N+1)/2 + (N+1)[x] \\ &= (x-[x])(x+1-[x])/2 + (x+1-[x])[x] \\ &= x(x+1)/2 + [x]/2 - [x]^2/2 \\ &= \tfrac{1}{2}\{x(x+1) - [x]([x]-1)\}. \end{aligned}$$

Since $y([x])$ is an arbitrary function of $[x]$ we may write finally $y(x+1) = x(x+1)/2 +$ an arbitrary function of $[x]$, the latter therefore being a periodic function of x of period unity. Accordingly $y(x) = x(x-1)/2 + A(x)$, just as if the summation had been effected for n instead of x but with a unit periodic in place of the arbitrary constant. In fact, of course, this example is merely a particular case of the solution we have already found for

$$y(x+1) - y(x) = P(x),$$

where $P(x)$ is a polynomial in x, the solution being expressed in terms of Bernoulli functions.

It is apparent now that the following statement is valid. If the solution of the equation

$$y_{n+1} - y_n = f(n)$$

is given by

$$y_n = A + F(n),$$

where $F(n) = \sum_{1}^{n-1} f(n)$, and A is an arbitrary constant, then the solution of the equation

$$y(x+1) - y(x) = f(x)$$

is given by

$$y(x) = B + F(x),$$

where B is an arbitrary unit periodic.

Example

$$y(x+1) - y(x) = e^x.$$

Since

$$\sum_{1}^{n-1} e^n = (e^n - e)/(e - 1),$$

the solution is

$$\begin{aligned} y(x) &= B + (e^x - e)/(e - 1) \\ &= C + e^x/(e - 1), \end{aligned}$$

where C is a unit periodic.

When the left-hand side of the difference equation can be thrown directly into the form of a complete difference the same procedure may be adopted.

Example

$$(x+1)y(x+1) - xy(x) = x.$$

Here

$$\Delta[xy(x)] = x,$$

and so $$xy(x) = x(x-1)/2 + A$$

or $$y(x) = (x-1)/2 + A/x.$$

EXERCISE (II)

By evaluating $\sum_{1}^{n-1} ne^n$, solve the equation

$$y(x+1) - y(x) = xe^x.$$

Where the right-hand side is not directly expressible in closed form when summed, it is sometimes convenient to express it as a convergent series.

Example

$$y(x+1) - y(x) = \frac{1}{x} = \sum_{n=0}^{\infty}\left(\frac{1}{x+n} - \frac{1}{x+n+1}\right)$$

$$= \sum_{n=0}^{\infty}\frac{1}{1+n}\left(\frac{x}{x+1+n} - \frac{x-1}{x+n}\right)$$

$$= f(x+1) - f(x),$$

where $f(x)$ is the convergent series

$$\sum_{n=0}^{\infty}\frac{x-1}{(1+n)(x+n)},$$

and $x > 0$. Accordingly a solution of the equation

$$\Delta y(x) = 1/x$$

is given by $$(x-1)\sum_{0}^{\infty}\frac{1}{(n+1)(n+x)}.$$

To this there must be added a unit periodic in x to provide the most general solution of

$$y(x+1) - y(x) = 1/x.$$

EXERCISES (III)

Solve the following difference equations—

1. $y(x+1) - y(x) = x^2$.
2. $xy(x+1) - (x+1)y(x) = x^2(x+1)$.
3. $y(x+1) - y(x) = 1/x(x+1)$.
4. $y(x+1) + y(x) = (-1)^x/x(x+1)$.

The Gamma Function

Just as the Bernoulli polynomials provide a system of relatively simple functions by means of which to express the solution of the difference equation

$$y(x + 1) - y(x) = P(x),$$

so the Gamma function provides the means whereby to express the solution of a difference equation of the form

$$y(x + 1)f(x) = y(x)g(x),$$

where $f(x)$ and $g(x)$ are given polynomials.

We turn therefore to a short examination of the Gamma function, $\Gamma(x)$, defined by means of the equation

$$y(x + 1) = xy(x)$$

one of the simplest equations of the foregoing type.

This equation defines the function of $y(x)$ with a degree of arbitrariness corresponding to a unit periodic, i.e. we can write

$$y(x) = A\Gamma(x),$$

where A is determined by the way in which $y(x)$ is initially specified over one single unit range.

Certain information regarding $\Gamma(x)$ is immediately available. For example, if x is any positive integer n

$$y(n + 1) = ny(n).$$

On solving the difference equation, we may take

$$\Gamma(n) = (n - 1)!.$$

Accordingly $\Gamma(x)$ is known at every positive integral value of x. Since $\Gamma(1) = 1$, and since $\Gamma(x)$ satisfies the equation—

$$\Gamma(x + 1) = x\Gamma(x),$$

it follows that

$$\lim_{x\to 0} x\Gamma(x) = 1.$$

Thus $\Gamma(x)$ becomes infinite like $1/x$ at $x = 0$. Moreover $\Gamma(x)$ has infinities of the same order for all negative integral values of x, for inserting $x = -1, -2, \ldots$

$$-\Gamma(-1) = \Gamma(0) \to \infty,$$

$$-2\Gamma(-2) = \Gamma(-1) \to \infty,$$

and so on.

It is to be expected therefore, if $\Gamma(x)$ could be expressed as an infinite product that $x,\ 1+\frac{x}{1},\ 1+\frac{x}{2}\ .\ .\ .$ would appear in the denominator. Of course

$$\begin{aligned}\Gamma(x) &= \Gamma(x+1)/x = \Gamma(x+2)/x(x+1)\\ &= \Gamma(x+3)/x(x+1)(x+2)\\ &= \Gamma(x+r)/x(x+1)(x+2)\ .\ .\ .\ (x+r-1)\\ &= \Gamma(x+r)/(r-1)!\,x(1+x/1)(1+x/2)\ .\ .\ .\ [1+x/(r-1)].\end{aligned}$$

Special Solution of the Equation

$$y(x+1) = xy(x).$$

Differentiating both sides logarithmically

$$\frac{y'(x+1)}{y(x+1)} - \frac{y'(x)}{y(x)} = \frac{1}{x}$$

or

$$z(x+1) - z(x) = 1/x.$$

We have already solved an equation of this type, the unknown function in this case being $y'(x)/y(x)$, so that we can write

$$\begin{aligned}z(x) = \frac{y'(x)}{y(x)} &= c + (x-1)\sum_0^\infty \frac{1}{(n+1)(n+x)}\\ &= c - \sum_0^\infty \left(\frac{1}{n+x} - \frac{1}{n+1}\right).\end{aligned}$$

We shall restrict ourselves to that solution of the original equation for which c is constant and not a unit periodic. This will certainly be the case when x is a positive integer, but that will not be the only case. The solution is then simply $(x-1)!$ as we have seen, and $y(1) = 1$. Now, however, integrating $y'(x)/y(x)$ and its expression, from 1 to $x+1$ we obtain

$$\log y(x+1) - \log y(1) = cx - \sum_{n=0}^\infty \left[\log \frac{x+n+1}{n+1} - \frac{x}{n+1}\right].$$

Thus

$$\log y(x+1) = \log \left[e^{cx} \Big/ \prod_0^\infty \left(1 + \frac{x}{n+1}\right) e^{-x/n+1}\right]$$

or $$y(x+1) = e^{cx}/\prod_1^\infty \left(1+\frac{x}{n}\right) e^{-x/n}.$$

We shall call the solution of the equation

$$y(x+1) = xy(x)$$

which is such that $y(1) = 1$ and c is a pure constant $\Gamma(x)$. Accordingly from what we have found—

$$\Gamma(x+1) = x\Gamma(x),$$

or $$\Gamma(x) = \Gamma(x+1)/x.$$

Hence $$\Gamma(x) = e^{cx}/x\prod_1^\infty \left(1+\frac{x}{n}\right) e^{-x/n},$$

and $$\Gamma(x+1) = e^{cx} \,.\, e^{c}/(x+1)\prod_1^\infty \left(1+\frac{x+1}{n}\right) e^{-(x+1)/n}.$$

Equating the two expressions for $\Gamma(x+1)$ we get

$$(x+1)e^{-c}\prod_1^\infty \left(1+\frac{x+1}{n}\right) e^{-(x+1)/n} = \prod_1^\infty \left(1+\frac{x}{n}\right) e^{-x/n},$$

or $$e^{-c} = \frac{1}{x+1}\prod_1^\infty \frac{1+\dfrac{x}{n}}{1+\dfrac{x+1}{n}} e^{1/n} = \lim_{n\to\infty} e^{\sum_1^n 1/n} /(n+1+x),$$

$$c = \lim_{n\to\infty}\left[-\sum_1^n \frac{1}{n} + \log n\right] = -\gamma.$$

Accordingly $$\Gamma(x) = \frac{e^{-\gamma x}}{x}\prod_{n=1}^\infty \frac{e^{x/n}}{(1+x/n)}.$$

EXERCISES (IV)

1. Show that the solution of

$$y(x+1) = x^r y(x),$$

is $$\{\Gamma(x)\}^r.$$

2. Show that if x is real, positive, and finite—

$$y(x) = \int_0^\infty e^{-t}\, t^{x-1}\, dt$$

satisfies the equation—

$$y(x+1) = xy(x),$$

and that $y(1) = 1$ and $y(\frac{1}{2}) = \sqrt{\pi}$.

This is Weierstrass' definition of the Gamma function. We proceed to examine yet another definition.

The Gamma Function—Euler's Definition

Let
$$u(z) = \frac{1}{z}\prod_{1}^{\infty}\frac{\left(1+\frac{1}{n}\right)^z}{1+\frac{z}{n}} \tag{1}$$

It is easy to show that this infinite product is convergent so that the expression on the right defines a definite function of z. Now

$$u(z+1) = \frac{1}{z+1}\prod_{n=1}^{\infty}\frac{\left(1+\frac{1}{n}\right)^{z+1}}{1+\frac{z+1}{n}} = \frac{1}{z+1}\prod_{n=1}^{\infty}\frac{\left(1+\frac{1}{n}\right)^{z}\cdot\frac{n+1}{n}}{\frac{n+z+1}{n}}$$

$$= \frac{1}{z+1}\prod_{n=1}^{\infty}\frac{\left(1+\frac{1}{n}\right)^{z}}{1+\frac{z}{n+1}} = \prod_{1}^{\infty}\frac{\left(1+\frac{1}{n}\right)^{z}}{1+\frac{z}{n}}$$

$$= zu(z).$$

Hence the function of z defined by (1) satisfies the difference equation

$$u(z+1) = zu(z). \tag{2}$$

From (1) it is obvious that

$$\lim_{z\to 0} zu(z) = 1.$$

Hence from (2), $$\Gamma(1) = 1.$$

Accordingly, if z is an integer $= r+2$

$$\Gamma(r+2) = \Gamma(1)(r+1)r(r-1)\ldots 1$$
$$= (r+1)!,$$

i.e.
$$\Gamma(r) = (r-1)! \tag{3}$$

where r is any positive integer.

We have seen that $\Gamma(z)$ satisfies a linear difference equation of the first order, viz. (2). We propose to show that $\Gamma(z)$ also satisfies the non-linear equation

$$\Gamma(z) \,.\, \Gamma(1 - z) = \pi/\sin \pi z. \tag{4}$$

We may derive this directly from the definition (1). For

$$\Gamma(z)\Gamma(1 - z) = \frac{1}{z(1 - z)}\prod_1^\infty \frac{\left(1 + \dfrac{1}{n}\right)^z}{1 + \dfrac{z}{n}} \cdot \frac{\left(1 + \dfrac{1}{n}\right)^{1-z}}{1 + \dfrac{1 - z}{n}}$$

$$= \frac{1}{z(1 - z)}\prod_1^\infty 1\Big/\left(1 + \frac{z}{n}\right)\left(1 - \frac{z}{n + 1}\right)$$

$$= \frac{1}{z(1 - z)}\Big/\left(1 + \frac{z}{1}\right)\left(1 - \frac{z}{2}\right)\left(1 + \frac{z}{2}\right)\left(1 - \frac{z}{3}\right)\ldots$$

$$= 1/z(1 - z^2)(1 - z^2/2^2)(1 - z^2/3^2)\ldots$$

$$= \pi/\sin \pi z.$$

Inserting $\quad z = \frac{1}{2},$

$$\Gamma(\tfrac{1}{2}) = \sqrt{\pi}.$$

Relation (4) establishes the connexion between the Γ function and the circular function. When $0 < z < \frac{1}{2}$ then $\frac{1}{2} < 1 - z < 1$.

Accordingly if $\Gamma(z)$ is tabulated over the range 0 to $\frac{1}{2}$, equation (4) enables it to be tabulated over the remainder of the range 0 to 1. Equation (2) then enables us to tabulate it over any positive range of z.

The known function $\Gamma(x)$ is therefore a solution of the difference equation

$$y(x + 1) = xy(x)$$

and its general solution is $A\Gamma(x)$ where A is an arbitrary unit periodic. Now consider—

1. $$y(x + 1) = (x - \alpha)y(x).$$

Let $\quad z = x - \alpha \qquad y(z + \alpha) = u(z),$

then $\quad u(z + 1) = zu(z),$

or $\quad u(z) = A\Gamma(z),$

i.e. $\quad y(z + \alpha) = A\Gamma(z)$

or $\quad y(x) = A\Gamma(x - \alpha).$

2. $$y(x+1) = (x-\alpha)(x-\beta)y(x).$$

Clearly $\Gamma(x-\alpha)\Gamma(x-\beta)$ is a solution of this equation since

$$(x-\alpha)\Gamma(x-\alpha)\,.\,(x-\beta)\Gamma(x-\beta) = \Gamma(x-\alpha+1)\,.\,\Gamma(x-\beta+1).$$

Thus the general solution of

$$y(x+1) = (x-\alpha)(x-\beta)y(x)$$

is
$$y(x) = A\Gamma(x-\alpha)\Gamma(x-\beta).$$

3. The general solution of

$$y(x+1) = (x-\alpha_1)(x-\alpha_2)\ldots(x-\alpha_r)y(x)$$

is
$$y(x) = A\Gamma(x-\alpha_1)\Gamma(x-\alpha_2)\ldots\Gamma(x-\alpha_r).$$

4. The general solution of

$$y(x+1) = c(x-\alpha)y(x)$$

is
$$y(x) = Ac^x\Gamma(x-\alpha).$$

This is obvious if we transform the original equation by writing

$$y(x) = c^x z(x),$$

when the equation reduces to that of (1).

5. The general solution of

$$y(x+1) = c(x-\alpha_1)(x-\alpha_2)\ldots(x-\alpha_r)y(x)$$

is
$$y(x) = Ac^x\Gamma(x-\alpha_1)\ldots\Gamma(x-\alpha_r).$$

6. Since any polynomial $f(x)$ of degree r can be written in the form

$$f(x) = c(x-\alpha_1)(x-\alpha_2)\ldots(x-\alpha_r)$$

where some of the factors may be identical, we can now write down the general solution of any equation of the form—

$$y(x+1) = f(x)y(x).$$

Examples

(i)
$$\begin{aligned} y(x+1) &= (2x^2-x-1)y(x) \\ &= (2x+1)(x-1)y(x) \\ &= 2(x+\tfrac{1}{2})(x-1)y(x). \end{aligned}$$

Hence
$$y(x) = A\,.\,2^x\Gamma(x+\tfrac{1}{2})\Gamma(x-1).$$

(ii)
$$\begin{aligned} y(x+1) &= (x - x^2)y(x) \\ &= x(1-x)y(x) \\ &= -x(x-1)y(x). \end{aligned}$$

Thus
$$\begin{aligned} y(x) &= A(-1)^x\Gamma(x)\Gamma(x-1) \\ &= A(-1)^x(x-1)\Gamma^2(x-1). \end{aligned}$$

7.
$$y(x+1) = \frac{f(x)}{g(x)}y(x)$$

where $f(x)$ and $g(x)$ are polynomials of degree r and s respectively. We may then write

$$f(x) = a\prod_{r=1}^{r}(x - \alpha_r)$$

$$g(x) = b\prod_{s=1}^{s}(x - \beta_s).$$

Then
$$y(x) = \left(\frac{a}{b}\right)^x \prod_{r=1}^{r}\Gamma(x - \alpha_r)/\prod_{s=1}^{s}\Gamma(x - \beta_s).$$

Another approach to the problem of solving linear difference equations is based on the Laplace transformation

$$y(z) = \int_{t_1}^{t_2} t^{z-1}f(t)\,dt.$$

The technique here is to find suitable values for t_1, t_2, and $f(t)$. The method can be applied to most linear difference equations with rational coefficients, but we shall illustrate the method for the moment by means of a first order equation. The examples we take actually concern $y(z)$ where z is the complex variable as the method applies equally well to such variables as it does to the real continuous variable x.

Example

$$y(z+1) = zy(z).$$

Let
$$y(z) = \int_{t_1}^{t_2} t^{z-1}f(t)\,dt,$$

then
$$\int_{t_1}^{t_2} t^z f(t)\,dt = z\int_{t_1}^{t_2} t^{z-1}f(t)\,dt = [t^z f(t)]_{t_1}^{t_2} - \int_{t_1}^{t_2} f'(t)t^z\,dt,$$

or
$$\int_{t_1}^{t_2} t^z[f(t) + f'(t)]\,dt = [t^z f(t)]_{t_1}^{t_2}.$$

Let $f(t) + f'(t) = 0$, then $f(t) = e^{-t}$, and if in addition we choose $t_2 = \infty$, $t_1 = 0$, and z to lie in the positive half-plane then

$$[t^z f(t)]_0^\infty \to 0$$

and
$$y(z) = \int_0^\infty t^{z-1} e^{-t}\, dt,$$

which is another of Euler's definitions of $\Gamma(z)$.

Derivation of a Second Order Equation from One of the First Order

It is obvious that since a first order equation is soluble in finite terms, a solution will be known for any second order equation derived from it by transformation. Consider

$$y(x+1) = \frac{a(x+\lambda)}{x+\mu}\, y(x),$$

which has a solution—

$$y(x) = a^x \Gamma(x+\lambda)/\Gamma(x+\mu).$$

We have
$$\Delta^n(x+\mu)y(x+1) = a\Delta^n(x+\lambda)y(x).$$

By Leibniz's theorem

$$(x+\mu)\Delta^n y(x+1) + n\Delta^{n-1}y(x+2)$$
$$= a[(x+\lambda)\Delta^n y(x) + n\Delta^{n-1}y(x+1)].$$

Let
$$z(x) = \Delta^{n-1}y(x),$$
then

$$(x+\mu)\Delta z(x+1) + nz(x+2) = a[(x+\lambda)\Delta z(x) + nz(x+1)],$$

or

$$(x+\mu)[z(x+2) - z(x+1)] + nz(x+2) - a(x+\lambda)z(x+1)$$
$$+ a(x+\lambda)z(x) - anz(x+1) = 0,$$

i.e.

$$(x+\mu+n)z(x+2)$$
$$- (x + ax + \mu + a\lambda + an)z(x+1) + a(x+\lambda)z(x) = 0.$$

A solution of this equation is therefore

$$z(x) = \Delta^{n-1}[a^x\Gamma(x+\lambda)/\Gamma(x+\mu)].$$

Example

$$(x + n)z(x + 2) - [(1 + a)x + an]z(x + 1) + axz(x) = 0.$$

A solution is $\quad z(x) = \Delta^{n-1}a^x = (a - 1)^{n-1}a^x.$

It is easily verified that a^x satisfies the equation. Show that this equation can be written

$$[(x + n)\mathrm{E} - x][\mathrm{E} - a]z(x) = 0,$$

and solve it completely.

How is one to recognize that the equation

$$(x + A)z(x + 2) - (Bx + C)z(x + 1) + a(x + \lambda)z(x) = 0$$

is of this type?

Identifying these two equations we have—

$$A = \mu + n, \qquad B = 1 + a, \qquad C = \mu + a\lambda + an.$$

These conditions demand, necessarily and sufficiently, that the sum of the linear coefficients associated with the terms $z(x + 2)$, $z(x + 1)$, $z(x)$ divided by $1—a$ must be a positive integer, irrespective of the value of x.

Example

$$(x + 1)z(x + 2) - (3x + 5)z(x + 1) + 2(x + 1)z(x) = 0.$$

Here the sum of the coefficients divided by $1 - 2$, i.e. -1, is 2 and this is the value of n. Hence a particular solution is

$$\Delta[2^x\Gamma(x + 1)/\Gamma(x - 1)] = \Delta[2^x x(x - 1)] = 2^x x(x + 3).$$

Verify and hence solve completely.

Special Classes of Linear Difference Equations, with Variable Coefficients, of the Second and Higher Orders

Success in the solving of difference equations, as with differential equations, usually depends on utilizing some special feature in the structure of the equation or some special piece of information that we may possess about the nature of the solution. We begin, therefore, with the treatment of certain special classes of equation or, what amounts to the same thing, the treatment of the general equations when some additional information is available regarding the nature of the solution; it is this additional information which it is hoped will point to the transformation or reconstruction of the equation so as to reduce our second order equation to dependence on one of the first order, or, in the case of equations of higher orders than two, to reduce the rth order equation to a dependence on one of order $(r - 1)$. This will enable us to dispose of a number of special forms before we turn to the general treatment of the equation and the derivation of solutions in the form of an expansion.

The general second order linear difference equation may be written

$$y(x + 2) + p(x)y(x + 1) + q(x)y(x) = R(x) \tag{5}$$

where $p(x)$, $q(x)$, and $R(x)$ are given functions of x. We call the equation

$$y(x + 2) + p(x)y(x + 1) + q(x)y(x) = 0 \tag{6}$$

which is the same as (5) but with zero on the right-hand side, the Complementary Equation. Clearly the complete solution of (5) will be of the form

$$y(x) = Au(x) + Bv(x) + S(x),$$

where A and B are arbitrary unit periodics, $u(x)$ and $v(x)$ are solutions of the complementary equation and $S(x)$ is a particular solution of the original equation (5).

Propositions

1. If a particular solution of the complementary equation (6) is known then the general solution of (5) containing the necessary two arbitrary constants, can be found.

Let $Y(x)$ be a known function which satisfies equation (6); thus we have

$$y(x + 2) + p(x)y(x + 1) + q(x)y(x) = R(x) \tag{5}$$

and

$$Y(x + 2) + p(x)Y(x + 1) + q(x)Y(x) = 0 \tag{7}$$

Writing

$$y(x) = Y(x)u(x),$$

where $u(x)$ is a new variable, (5) becomes

$$Y(x + 2)u(x + 2) + p(x)Y(x + 1)u(x + 1) + q(x)Y(x)u(x) = R(x).$$

Multiplying (7) by $u(x + 1)$ and subtracting from the above equation we obtain—

$$Y(x + 2)\Delta u(x + 1) - q(x)Y(x)\Delta u(x) = R(x) \tag{8}$$

which is a first order equation in $\Delta u(x)$ from which a formal expression for $\Delta u(x)$ may be obtained. An expression for $u(x)$ and so for $y(x)$ follows immediately.

Note that (8) may be written

$$Y(x+2)\left[\frac{y(x+2)}{Y(x+2)}-\frac{y(x+1)}{Y(x+1)}\right]-q(x)Y(x)\left[\frac{y(x+1)}{Y(x+1)}-\frac{y(x)}{Y(x)}\right] = R(x)$$

or

$$Y(x+1)y(x+2) - Y(x+2)y(x+1) - q(x)Y(x)y(x+1) + q(x)y(x)Y(x+1) = R(x)Y(x+1)$$

or

$$[\mathrm{E} - q(x)][Y(x)y(x+1) - Y(x+1)y(x)] = R(x)Y(x+1). \quad (9)$$

We see therefore that a knowledge of a particular solution $Y(x)$ of the complementary equation (6) enables us to factorize the left-hand side of the original equation (5) as in the form (9) by multiplying this original equation throughout by $Y(x+1)$.

From a different angle we may say $Y(x+1)$ is a factorizing function of equation (5). We must bear in mind that since the coefficients of these operators in (9) are functions of x the operators are not commutative so that the order of their appearance on the left cannot be altered.

Thus, given a second order linear difference equation and some known function $Y(x)$ satisfying the complementary equation (6), the solution of the original second order equation can be effected through the first order equation (8).

Example

$$(x-1)y(x+2) + (2-3x)y(x+1) + 2xy(x) = 0.$$

To find the solution of this equation for $x \geqslant 0$, given that a particular solution is $y(x) = x$, and that $y(x) = \sin \pi x$ for $0 \leqslant x < 2$. Thus

$$(x-1)(x+2) + (2-3x)(x+1) + 2x^2 = 0,$$

and by eliminating the middle term, we obtain

$$(x-1)[(x+1)y(x+2) - (x+2)y(x+1)] + 2x[(x+1)y(x) - xy(x+1)] = 0,$$

or $$\Delta\{[xy(x+1) - (x+1)y(x)]/2^{x-1}(x-1)\} = 0;$$

hence, $$xy(x+1) - (x+1)y(x) = A(x) \, . \, 2^{x-1}(x-1)$$

where $A(x)$ is a unit periodic. Thus,

$$\Delta[y(x)/x] = A(x)2^{x-1}(x-1)/x(x-1).$$

Splitting the right-hand side into partial fractions we obtain

$$\Delta[y(x)/x] = A(x)2^{x-1}\left[\frac{2}{x+1} - \frac{1}{x}\right]$$

$$= \frac{A(x+1)2^x}{x+1} - \frac{A(x)2^{x-1}}{x},$$

since $$A(x) = A(x+1).$$

Hence $$y(x) = A(x)2^x + xB(x)$$

where $A(x)$ and $B(x)$ are periodic functions of period unity.

We now apply the boundary conditions to determine $A(x)$ and $B(x)$, the unit periodics.

$$\sin \pi x = A(x)2^x + xB(x) \quad \text{for } 0 \leqslant x < 2$$

or $$\sin \pi[x] = A([x])2^{[x]} + [x]B([x]) \quad \text{for } 0 \leqslant x < 1,$$

and $\sin \pi(1+[x]) = A([x])2^{1+[x]} + (1+[x])B([x])$ for $1 \leqslant x < 2$,

where $[x] = x - N$, where N is the greatest integer occurring in x. Hence

$$A([x]) = \frac{\sin \pi[x] \, . \, (1+2[x])}{2^{[x]}(1-[x])} = A(x)$$

and $$B([x]) = \frac{-3 \sin \pi[x]}{1-[x]} = B(x),$$

since $A(x)$ and $B(x)$ are unit periodics. Thus

$$y(x) = \frac{2^N(1+2x-2N) - 3x}{1-x+N} \, . \sin \pi(x-N)$$

$$= (-1)^N \left\{\frac{(1-2N+2x)2^N - 3x}{1+N-x}\right\} \sin \pi x$$

for any positive value of x.

EXERCISES (V)

1. Show that if

$$p(x) + q(x) + 1 = 0,$$

then the equation

$$y(x+2)+p(x)y(x+1)+q(x)y(x)=R(x)$$

is in fact an equation of the first order and may be written as

$$[\mathrm{E}-q(x)]\Delta y(x)=R(x).$$

2. By extending the method given above show that if any particular solution of the equation

$$y(x+r)+p_1y(x+r-1)+p_2y(x+r-2)+\ldots+p_ry(x)=0$$

is known, then the order of the difference equation

$$y(x+r)+p_1y(x+r-1)+p_2y(x+r-2)+\ldots+p_ry(x)=R(x)$$

may be reduced by unity.

[$p_1, p_2, \ldots, p_r$, $R(x)$ are all given functions of x.]

3. Solve the equation—

$$(x-1)y(x+2)+(2-3x)y(x+1)+2xy(x)=1.$$

4. Obtain formal solutions of the following equations—

(i) $y_{n+2}+(n^2+1)y_{n+1}+n^2y_n=0.$

(ii) $ny_{n+2}+(n+1)y_{n+1}-a(na+n+1)y_n=0.$

(iii) $ny_{n+2}-(n+1)y_{n+1}+y_n=n.$

(iv) $y_{n+2}+(n+1)y_{n+1}+(3n-6)y_n=0.$

(v) $y_{n+2}+(n+1)y_{n+1}+ny_n=0.$

(vi) $ny_{n+2}-4ny_{n+1}+(3n+6)y_n=0.$

6. If the equations

$$y_{n+2}+p_ny_{n+1}+q_ny_n=0$$

and

$$z_{n+2}+P_nz_{n+1}+Q_nz_n=0$$

have a solution in common, show that

$$(p_n-P_n)(p_{n-1}Q_{n-1}-P_{n-1}q_{n-1})=(Q_{n-1}-q_{n-1})(q_n-Q_n).$$

7. Solve the equation for y_n,

$$ny_{n+2}+n^2y_{n+1}-(n^2+2n+2)y_n=0,$$

$$nz_{n+2}+n^3z_{n+1}-(2+n+n^2+n^3)z_n=0,$$

given that these have a solution in common.

8. Transform the equation

$$y(x+2)+p(x)y(x+1)+q(x)y(x)=0$$

to the form

$$z(x+2)+z(x+1)+Q(x)z(x)=0,$$

by means of the transformation

$$z(x)=R(x)y(x).$$

[$p(x)$, $q(x)$ are known, $Q(x)$, $R(x)$ to be found.]

Similarly, change the form of

$$y(x + 2) + p(x)y(x + 1) + q(x)y(x)$$

to

$$z(x + 2) + P(x)z(x + 1) + z(x) = 0.$$

With this preliminary introduction we may now turn to a consideration of the general linear equation from the point of view of any special structural features it may show. What we have really done so far has been to use a particular solution, when it was known, as a means to bring out this structure and so to enable the complete solution to be found. We now examine these linear equations that are capable of reduction to simpler forms by operational factorization.

2. If the equation

$$y(x + r) + p_1 y(x + r - 1) + \ldots + p_r y(x) = R(x) \tag{10}$$

(where $p_1, p_2, \ldots, p_r$ are functions of x) is operationally factorizable in the form

$$(a_1\mathrm{E} + b_1)(a_2\mathrm{E} + b_2) \ldots (a_r\mathrm{E} + b_r)y(x) = R(x)$$

then the general solution of (10) can be found by solving r linear difference equations each in one dependant variable. Thus, let

$$(a_2\mathrm{E} + b_2) \ldots (a_r\mathrm{E} + b_r)y(x) = z(x)$$

then

$$(a_1\mathrm{E} + b_1)z(x) = R(x).$$

From this $z(x)$ immediately follows. Now let

$$(a_3\mathrm{E} + b_3) \ldots (a_r\mathrm{E} + b_r)y(x) = w(x)$$

then

$$(a_2\mathrm{E} + b_2)w(x) = z(x)$$

from which $w(x)$ follows immediately. By following this procedure systematically we finally arrive at the expression for $y(x)$ with the appropriate number of arbitrary constants.

EXERCISES (VI)

1. Solve—

(i) $(\mathrm{E} + x)(\mathrm{E} - x)y(x) = 0.$
(ii) $(\mathrm{E} - x)^2 y(x) = 0.$

2. Determine in what circumstances

$$[\mathrm{E} - p(x)][\mathrm{E} - q(x)]y(x) \equiv [\mathrm{E} - q(x)][\mathrm{E} - p(x)]y(x).$$

Assuming that the conditions are satisfied, write down the solution of the equation—

$$y(x + 2) - [p(x) + q(x + 1)]y(x + 1) + p(x)q(x)y(x) = 0.$$

3. Solve—

(i) $y(x+2) - [a(x) + a(x+1)]y(x+1) + a^2(x)y(x) = 0$.
(ii) $xy(x+2) - (x^2 + x + 1)y(x+1) + xy(x) = 0$.
(iii) $y(x+2) + 2\lambda^x y(x+1) + \lambda^{2x-1}y(x) = 0$.
(iv) $y(x+2) + (2x+1)y(x+1) + x^2y(x) = 0$.

Thus we see that the problem of solving second and higher order equations becomes that of finding the operational factors (if they exist in the original form of the equation), or finding some multiplying function which will enable the resulting equation to be factorized.

3. Consider as before the second order equation

$$y(x+2) + p(x)y(x+1) + q(x)y(x) = R(x). \tag{5}$$

In section 1 we saw that a knowledge of a particular solution $Y(x)$ of the complementary equation, (6) enables us to carry out the transformation $y(x) = Y(x)u(x)$ and so reduce the second order equation to one of the first order in $\Delta u(x)$ which can be written in the form

$$[\mathrm{E} - q(x)][Y(x)y(x+1) - Y(x+1)y(x)] = R(x)Y(x+1). \tag{9}$$

This suggests a possible approach towards the solution of the equation (5) in cases where $Y(x)$ is not known.

Let
$$\begin{aligned} u(x) &= b(x)y(x+1) + c(x)y(x) \\ &= [b(x)\mathrm{E} + c(x)]y(x) \end{aligned} \tag{11}$$

where $b(x)$ and $c(x)$ are, for the moment, unspecified functions of x.

Then, $u(x+1) = b(x+1)y(x+2) + c(x+1)y(x+1)$.

Thus
$$\begin{aligned} u(x+1) + Q(x)u(x) &= b(x+1)y(x+2) \\ &+ [b(x)Q(x) + c(x+1)]y(x+1) + Q(x)c(x)y(x), \end{aligned} \tag{12}$$

where $Q(x)$ is again some unspecified function of x. Now, if we seek to write the left-hand side of (5) proportional to the right-hand side of (12) then,

$$b(x+1) = \frac{b(x)Q(x) + c(x+1)}{p(x)} = \frac{Q(x)c(x)}{q(x)},$$

or
$$b(x)Q(x) + c(x+1) = b(x+1)p(x), \tag{13}$$

and
$$b(x+1)q(x) = c(x)Q(x). \tag{14}$$

Thus from (12) and (5), we have

$$[\mathrm{E} + Q(x)]u(x) = R(x)b(x + 1),$$

or
$$[\mathrm{E} + Q(x)][b(x)\mathrm{E} + c(x)]y(x) = R(x)b(x + 1),$$

on using the expression for $u(x)$ in (11). This is in fact the original equation thrown into factors on multiplication throughout by the function $b(x + 1)$. To determine $b(x)$ we notice that (13) and (14) can be combined to form the equation

$$q(x + 1)b(x + 2) - p(x)Q(x + 1)b(x + 1) + Q(x)Q(x + 1)b(x) = 0, \quad (15)$$

where $p(x)$ and $q(x)$ are given functions of x and $b(x)$ and $Q(x)$ are as yet unspecified.

Thus, if we choose $Q(x)$ arbitrarily we must solve an equation of the second order to find $b(x)$; on the other hand, if we choose $b(x)$, the equation in $Q(x)$ seems at first sight to be of the first order. It is so, in fact, in so far as its complete solution must contain one arbitrary constant, but, as can easily be seen, in order to arrive at that solution it is again necessary to solve a linear equation of the second order. For, writing

$$Q(x) = P(x)/P(x + 1)$$

equation (15) becomes—

$$b(x + 2)q(x)P(x + 2) - b(x + 1)p(x)P(x + 1) + b(x)P(x) = 0.$$

Whichever way we look at this problem we cannot apparently evade finding a solution of some second order linear equation. At the same time we have widened the possibility of solution by showing that there are sets of linear equations whose solutions are dependent on one another, the solution of one being a factorizing function of another.

So far the functions $b(x)$ and $Q(x)$ have been arbitrary. We now proceed to discuss various choices for the function $Q(x)$, and investigate the resulting equation in $b(x)$ and the consequent resulting factorization of the original equation in $y(x)$, viz. (5).

A. In the particular case where

$$Q(x) = - q(x),$$

from (14),
$$b(x + 1) = - c(x)$$

and substituting in (13) we obtain

$$b(x+2)+p(x)b(x+1)+q(x)b(x)=0,$$

which is, of course, the complementary equation, (6).

B. If we wish to find a multiplier that will make the left-hand side of (5) a complete difference, that is if we wish to display the factor (E − 1), then we must make $Q(x) = -1$. In that case, from (14)

$$b(x+1)q(x) = -c(x)$$

and, substituting in (13),

$$q(x+1)b(x+2)+p(x)b(x+1)+b(x)=0. \tag{16}$$

This is a second order equation in $b(x)$; and $b(x+1)$ is a summating (or integrating) factor of the original equation as, if some function $b(x)$ be known, (5) may be written—

$$\Delta[b(x)\mathrm{E}-q(x)b(x+1)]y(x)=R(x)b(x+1). \tag{17}$$

We may call (16) the adjoint equation of (5). It may be rewritten as

$$b(x+2)+\frac{p(x)}{q(x+1)}\,.\,b(x+1)+\frac{1}{q(x+1)}\,.\,b(x)=0.$$

Again, this adjoint equation has in its turn, a summating factor, say, $B(x+1)$, where $B(x)$ satisfies the equation

$$B(x+2)+B(x+1)p(x)q(x+2)/q(x+1)+B(x)q(x+2)=0 \tag{18}$$

and equation (18) in $B(x)$ has in its turn a summating factor $C(x+1)$ say, where $C(x)$ satisfies the equation

$$C(x+2)+C(x+1)\frac{p(x)q(x+2)}{q(x+1)q(x+2)}+\frac{C(x)}{q(x+3)}=0 \tag{19}$$

and so on.

This provides a set of equations the solution of each of which offers a summating factor for the previous one of the set. If any one of these equations can be solved in closed form, then every member of the set can be solved. We notice that $B(x) = q(x)y(x)$ satisfies (18) since, on substitution, we obtain

$$q(x+2)y(x+2)+p(x)q(x+2)y(x+1)+q(x+2)y(x)=0,$$

which is, of course, simply the original equation (5). By the same argument we have

$$C(x) = b(x)/q(x+1),$$

and again, at the next stage

$$D(x) = q(x+2)B(x) = q(x+2)q(x)y(x),$$

and so on.

This can clearly be extended to the general linear difference equation of the rth order, viz.

$$y(x+r) + p_1(x)y(x+r-1) + \dots + p_r(x)y(x) = R(x)$$

where it can be shown that this equation can be written as the complete difference of an expression of order $(r-1)$ by multiplying throughout by a function obtained from the solution of another adjoint equation of the rth order. In the general case, also, as in that of the second order, there exists sets of adjoint equations. If any one is capable of solution in closed form, then every member of the set is capable of such solution also.

Corollary

The necessary and sufficient condition for

$$y(x+2) + p(x)y(x+1) + q(x)y(x) = R(x)$$

to be exact as it stands is

$$p(x-1) + q(x) + 1 = 0.$$

That it is necessary follows immediately from the statement that if the equation is exact a solution of (16) is $b(x) = 1$. That it is sufficient may be seen by a direct consideration of the equation when this condition is imposed upon it. Substituting for $q(x)$, we obtain,

$$y(x+2) + p(x)y(x+1) - (1 + p(x-1))y(x) = R(x),$$

i.e. $$y(x+2) - y(x+1) + y(x+1) - y(x) + p(x)y(x+1) - p(x-1)y(x) = R(x),$$

i.e. $$\Delta\{y(x+1) + y(x) + p(x-1)y(x)\} = R(x).$$

Example

Consider the equation

$$2x(x-1)y(x+2) - (x-1)(x+2)y(x+1) + xy(x) = 0.$$

The adjoint equation is

$$b(x+2) - (x+2)b(x+1) + 2xb(x) = 0.$$

Let $$b(x) = a^x u(x),$$

then $$a^2 u(x+2) - a(x+2)u(x+1) + 2xu(x) = 0.$$

If $a = 2$ this gives $4u(x+2) - 2(x+2)u(x+1) + 2xu(x) = 0$

and $u(x) = 1$ is a solution. Thus, $b(x) = 2^x$ is a solution of the adjoint equation, and $b(x+1) = 2^{x+1}$ is a summating factor of the original equation, which becomes

$$\Delta\left[2^x y(x+1) - \frac{2^{x+1}}{2\,.\,(x-1)}\,y(x)\right] = 0.$$

Hence, $$y(x+1) - \frac{y(x)}{(x-1)} = A(x)\,.\,2^{-x},$$

where $A(x)$ is an arbitrary unit periodic. Multiplying through by $(x-1)\Gamma(x-1)$ which equals $\Gamma(x)$,

$$\Delta[\Gamma(x-1)y(x)] = A(x)\Gamma(x)\,.\,2^{-x}$$

$$\Gamma(x-1)y(x) = A(x)\sum_{N=0}^{N-1}\Gamma\,([x]+N)\,.\,2^{-[x]-N} + B(x)$$

where N is the greatest integer occurring in x and $A(x)$ and $B(x)$ are arbitrary unit periodics.

EXERCISES (VII)

1. Solve the following equations—

(i) $y(x+2) - (x+2)^2 y(x+1) + x(x+1)(x+2)y(x) = 0.$

(ii) $y(x+2) - \dfrac{y(x)}{(x+1)(x+2)} = 0.$

(iii) $ay(x+2) + (a^x+1)y(x+1) + a^{x-1}y(x) = 0, 0 < a < 1.$

(iv) $xy(x+2) + (x+1)(x^2+x-1)y(x+1) - x^2(x+1)y(x) = 0.$

(v) $y(x+2) - e^x(e-1)y(x+1) - e^{2x}y(x) = 0.$

(vi) $y(x+2) + y(x+1) - \left(\dfrac{2x-1}{x+1}\right)y(x) = 0.$

2. Show that

(i) $y_{n+2} + ny_{n+1} + (n-3)y_n$

becomes a complete difference on multiplying it by $(-1)^n(n-3)$.

Hence show that $(-1)^n(n-4)(n-4)!$ is a particular solution of the equation

$$y_{n+2} + ny_{n+1} + (n-3)y_n = 0.$$

(ii) $4y_{n+2} + 8ny_{n+1} - (4n - 7)y_n = 0$

can be solved by a summating factor $(2n - 1)2^{n-1}$.

(iii) 3^n is a summating factor of

$$3(n + 1)y_{n+2} - n\,(n + 2)y_{n+1} + \frac{n(n - 1)}{3}\,y_n = 0,$$

and hence solve the equation.

(iv) We may determine a factor $ax + b$ which when multiplied into

$$y(x + 2) + (x + 1)y(x + 1) - xy(x)$$

converts it into a complete difference.

(v) $9xy(x + 2) - (x + 1)y(x + 1) + (2x + 1)y(x) = 0$

is made exact by multiplying by a factor of the form $\lambda^x(ax + b)$. Hence solve the equation.

(vi) If $y_n = y_0 + n\Delta y_0 + \dfrac{n(n - 1)}{2\,!}\,\Delta^3 y_0$

then
$$y_n = y_0 + n\Delta y_n - \frac{n(n + 1)}{2\,!}\,\Delta^2 y_n.$$

Solve this second order equation completely.

3. Show that if the solution of the equation

$$y(x + 2) + p(x)y(x + 1) + q(x)y(x) = 0$$

is known, then the solutions of all equations of the form

$$z(x + 2) + P(x)z(x + 1) + Q(x)z(x) = 0$$

are also known, where

$$p(x)p(x - 1)/q(x) = P(x)P(x - 1)/Q(x).$$

4. Solve the equations

$$y(x + 2) + ap(x)y(x + 1) + p(x)p(x - 1)y(x) = 0,$$

and
$$y(x + 2) + xy(x + 1) + x(x - 1)y(x) = 0,$$

by transforming them into second order linear equations with constant coefficients.

5. Find the function $\phi(x)$, $\{z(x) = \phi(x)y(x)\}$, that transforms the equation

$$y(x + 2) + p(x)y(x + 1) + \frac{p(x)p(x - 1)}{x + 1}\,y(x) = 0$$

into
$$z(x + 2) + (x + 1)z(x + 1) + xz(x) = 0.$$

Hence solve, $y(x + 2) + (x + 1)^2y(x + 1) + x^2(x + 1)y(x) = 0$.

We turn now to an examination of the procedure to be adopted when some information is available concerning the nature of the solutions irrespective of whether or not the structures of the equations themselves suggest some line of approach.

Let us suppose we are concerned with the solution of the second order linear difference equation—

$$y(x + 2) + p(x)y(x + 1) + q(x)y(x) = 0. \tag{6}$$

Let $u(x)$ be a solution which is as yet unknown, and $v(x)$ is another solution and is a known function of x, $u(x)$, etc.

Then $\quad v(x) = f[x, u(x), u(x - 1), \ldots, u(x - r)]$

or $\quad v(x + r) = f[x + r, u(x + r), u(x + r - 1), \ldots, u(x)].$

Thus we have the two equations

$$u(x + 2) + p(x)u(x + 1) + q(x)u(x) = 0 \tag{19}$$

and

$$\begin{aligned} f[x + r + 2, u(x + r + 2), \ldots, u(x + 2)] \\ + p(x + r)f[x + r + 1, u(x + r + 1), \ldots, u(x + 1)] \\ + q(x + r)f[x + r, u(x + r), \ldots, u(x)] = 0. \end{aligned} \tag{20}$$

The first of these, (19) enables us to express $u(x + r + 2)$ in terms of $u(x)$ and $u(x + 1)$, for $r \geqslant 0$; when these expressions are inserted in the second equation we are accordingly left with an expression of the form

$$F[x, u(x), u(x + 1)] = 0$$

a difference equation of the first order from which to determine $u(x)$. If, therefore, this is soluble in finite terms then so also is the original equation (6) and so, theoretically, would an equation of the type (5), viz.

$$y(x + 2) + p(x)y(x + 1) + q(x)y(x) = R(x) \tag{5}$$

be soluble.

We illustrate with four special forms:

1. One solution is $\phi(x)$ times the other, where $\phi(x)$ is known. Then, if $u(x)$ and $\phi(x)u(x)$ are the two solutions, we have

$$u(x + 2) + p(x)u(x + 1) + q(x)u(x) = 0 \tag{19}$$

and

$$\phi(x + 2)u(x + 2) + p(x)\phi(x + 1)u(x + 1) + q(x)\phi(x)u(x) = 0.$$

Eliminating the terms in $u(x + 2)$, we obtain

$$\begin{aligned} p(x)[\phi(x + 2) - \phi(x + 1)]u(x + 1) \\ + q(x)[\phi(x + 2) - \phi(x)]u(x) = 0 \end{aligned}$$

a first order equation from which we may find $u(x)$.

Example

$$ny_{n+2} - (n + 2)(n^2 + 3n + 1)y_{n+1} + (n + 1)^3(n + 2)y_n = 0.$$

It is known that one solution is $n!$ times the other. The equation for u_n becomes

$$-\frac{(n + 2)(n^2 + 3n + 1)}{n}[(n + 2)! - (n + 1)!]u_{n+1}$$
$$+ \frac{(n + 1)^3(n + 2)}{n}[(n + 2)! - n!]u_n = 0,$$

or $(n^2 + 3n + 1)[(n + 2)(n + 1) - n + 1]u_{n+1}$
$$- (n + 1)^3[(n + 2)(n + 1) - 1]u_n = 0,$$

or

$$(n^2 + 3n + 1)(n^2 + 2n + 1)u_{n+1} - (n + 1)^3(n^2 + 3n + 1)u_n = 0,$$

or $$u_{n+1} - (n + 1)u_n = 0.$$

Hence $u_n = n!$. Thus the solution of the equation is

$$y_n = An! + B(n!)^2.$$

2. Let it be known that the difference equation

$$y(x + 2) + p(x)y(x + 1) + q(x)y(x) = 0 \tag{6}$$

is such that one solution is the square of the other.

Then $$u(x + 2) + p(x)u(x + 1) + q(x)u(x) = 0$$
and $$u^2(x + 2) + p(x)u^2(x + 1) + q(x)u^2(x) = 0$$

where $u(x)$ is the particular solution. Hence

$$[p(x)u(x + 1) + q(x)u(x)]^2 + p(x)u^2(x + 1) + q(x)u^2(x) = 0,$$

or $p(x)[1 + p(x)]u^2(x + 1)$
$$+ 2p(x)q(x)u(x + 1)u(x) + q(x)[1 + q(x)]u^2(x) = 0.$$

This is a homogeneous equation which can be solved for $u(x + 1)/u(x)$ giving

$$\frac{u(x + 1)}{u(x)} = \frac{-p(x)q(x) \pm \sqrt{[p^2(x)q^2(x) - p(x)q(x)(1 + p(x))(1 + q(x))]}}{p(x)[1 + p(x)]}$$
$$= \frac{-p(x)q(x) \pm \sqrt{[-p(x)q(x)(1 + p(x) + q(x))]}}{p(x)[1 + p(x)]},$$

i.e. a first order equation which can be solved for the particular solution $u(x)$.

Example

It is known that one solution of the equation

$$xy(x+2) - (x+2)(x^2+3x+1)y(x+1) + (x+1)^3(x+2)y(x) = 0$$

is the square of another. Here

$$p(x) = -(x+2)(x^2+3x+1)/x,$$

and

$$q(x) = (x+1)^3(x+2)/x.$$

Thus, it can be seen that either

$$\frac{u(x+1)}{u(x)} = (x+1),$$

or

$$\frac{u(x+1)}{u(x)} = \frac{(x+1)(x^2+2x+2)}{x^2+4x+2}.$$

If these two alternatives are substituted back into the equation it will be found that

$$\frac{u(x+1)}{u(x)} = (x+1)$$

satisfies the equation and the other solution does not. Hence, a solution of the original equation is

$$u(x) = \Gamma(x+1)$$

and another solution is

$$u(x) = [\Gamma(x+1)]^2.$$

Hence, in general

$$y(x) = A\Gamma(x) + B[\Gamma(x+1)]^2$$

where A and B are arbitrary unit periodics.

3. Suppose one solution of the difference equation

$$y(x+2) + p(x)y(x+1) + q(x)y(x) = 0$$

is known to be the inverse of another solution. Then

$$u(x+2) + p(x)u(x+1) + q(x)u(x) = 0,$$

and

$$\frac{1}{u(x+2)} + \frac{p(x)}{u(x+1)} + \frac{q(x)}{u(x)} = 0.$$

Hence $$[p(x)u(x+1)+q(x)u(x)]\left[\frac{p(x)}{u(x+1)}+\frac{q(x)}{u(x)}\right]=1,$$

or $$p^2(x)+q^2(x)+p(x)q(x)\left[\frac{u(x)}{u(x+1)}+\frac{u(x+1)}{u(x)}\right]=1.$$

From this $u(x+1)/u(x)$ follows. As before, only one of the two values for $\dfrac{u(x+1)}{u(x)}$ satisfies the original difference equation.

EXERCISE (VIII)

$$x^2(x+2)y(x+2)-(2x+1)(x^2+x+1)y(x+1)$$
$$+(x+1)(x^2-1)y(x)=0.$$

It is known that the above equation has two solutions whose product is unity.

Show that the general solution of the difference equation may be written

$$y(x)=C_1x(x-1)+\frac{C_2}{x(x-1)}$$

where C_1 and C_2 are arbitrary functions of period unity.

4. If one solution of the equation

$$y(x+2)+p(x)y(x+1)+q(x)y(x)=0$$

is $u(x)$ and another is $u(x-1)$, it follows that—

$$u(x+2)+p(x)u(x+1)+q(x)u(x)=0$$

and $$u(x+1)+p(x)u(x)+q(x)u(x-1)=0$$

or $$u(x+2)+p(x+1)u(x+1)+q(x+1)u(x)=0.$$

Hence $$\frac{u(x+1)}{u(x)}=-\frac{\Delta q(x)}{\Delta p(x)},$$

from which we obtain $u(x)$ and hence the general solution of the equation.

The relation which must be satisfied by $p(x)$ and $q(x)$ is

$$-\frac{\Delta q(x+1)}{\Delta p(x+1)}+p(x)-q(x)\frac{\Delta p(x)}{\Delta q(x)}=0$$

or $$p(x)q(x+1)-q(x)p(x+1)=\frac{\Delta q(x+1)\Delta q(x)}{\Delta p(x+1)}.$$

Example

$$y(x+2) - 2(x+1)y(x+1) + x(x+1)y(x) = 0.$$

It is easily verified that the above relation between the coefficients is satisfied in this case. Thus

$$u(x+1) = (x+1)u(x);$$

therefore $$u(x) = \Gamma(x+1),$$

and we have $$y(x) = A(x)\Gamma(x+1) + B(x)\Gamma(x).$$

EXERCISES (IX)

1. Given that there exist two solutions, $y(x)$ and $x^2y(x)$, of the equation

$$(2x+1)y(x+2) - 8(x+1)y(x+1) + 4(2x+3)y(x) = 0,$$

solve the equation.

2. Solve the equation

$$(n+2)(2n+1)y_{n+2} - 4(n+1)^2y_{n+1} + n(2n+3)y_n = 0$$

where the product of two solutions is unity.

3. Show that if one solution $v(x)$ of a second order equation is a given function of another solution of the form

$$v(x) = f[x, u(x), u(x-1), \ldots, u(x-r)]$$

then the second order equation can be made dependent for its solution on an equation of the first order.

Work out in detail the case where

$$v(x) = 1/u(x).$$

Solution of Linear Finite Difference Equations by Means of Definite Integrals

We illustrate first with a special case. Our purpose is to obtain a particular solution by means of which, of course, the equation may be generally solved. The particular solution will be found in the form of a definite integral containing the independent variable. Consider

$$y(z+2) - (z-1)y(z+1) - zy(z) = 0.$$

Let $$y(z) = \int_{t_1}^{t_2} t^{z-1}f(t)\,dt$$

where $f(t)$, t_1 and t_2 have to be determined. Inserting this expression into the equation we have

$$\int_{t_1}^{t_2} t^{z+1} f(t)\,dt - (z-1)\int_{t_1}^{t_2} t^z f(t)\,dt - z\int_{t_1}^{t_2} t^{z-1} f(t)\,dt = 0,$$

or

$$\int_{t_1}^{t_2} t^{z-1}(t^2+t)f(t)\,dt - z\int_{t_1}^{t_2} t^{z-1}(t+1)f(t)\,dt = 0.$$

Integrating the second integral by parts leads to

$$z\int_{t_1}^{t_2} t^{z-1}(t+1)f(t)\,dt = [(t+1)f(t)t^z]_{t_1}^{t_2} - \int_{t_1}^{t_2} t^z \frac{d}{dt}[(t+1)f(t)]\,dt.$$

Hence

$$\int_{t_1}^{t_2} t^{z-1}\left[(t^2+t)f(t) + t\frac{d}{dt}\{(t+1)f(t)\}\right]dt - [(t+1)f(t)t^z]_{t_1}^{t_2} = 0.$$

Accordingly, let $f(t)$ satisfy the differential equation—

$$t(t+1)f(t) + t\frac{d}{dt}\{(t+1)f(t)\} = 0,$$

and t_2 and t_1 be chosen to make

$$[(t+1)f(t)t^z]_{t_1}^{t_2} = 0.$$

The former of these is satisfied by

$$(t+1)f(t) = \mathrm{e}^{-t} \text{ or } f(t) = \mathrm{e}^{-t}/(t+1),$$

and the latter requires

$$[\mathrm{e}^{-t}\, t^z]_{t_1}^{t_2} = 0.$$

It follows that if $t_2 = \infty$ and $t_1 = 0$ *z may lie anywhere in the positive half plane*. Thus a solution of the equation is

$$y(z) = \int_0^\infty dt\, t^{z-1}\,\mathrm{e}^{-t}/(t+1).$$

Verification

On inserting into the original equation, we have

$$\int_0^\infty dt\,\frac{t^{z+1}\,\mathrm{e}^{-t} - (z-1)t^z\,\mathrm{e}^{-t} - zt^{z-1}\,\mathrm{e}^{-t}}{t+1}$$

$$= \int_0^\infty \frac{dt}{(t+1)}[t^z(t+1) - zt^{z-1}(t+1)]\,\mathrm{e}^{-t}$$

$$= \int_0^\infty dt\, e^{-t}(t^z - zt^{z-1}) = \Gamma(z+1) - z\Gamma(z) = 0.$$

Note

The original equation can be written

$$[y(z+2) + y(z+1)] - z[y(z+1) + y(z)] = 0.$$

Hence $$y(z+1) + y(z) = A\Gamma(z)$$

and the general solution follows at once.

Consider the more general form

$$y(z+2) + (az+b)y(z+1) + (cz+d)y(z) = 0. \tag{21}$$

Let $$y(z) = \int_{t_1}^{t_2} t^{z-1} f(t)\, dt \tag{22}$$

and insert this into the difference equation. Hence

$$\int_{t_1}^{t_2} (t^{z+1} + bt^z + dt^{z-1}) f(t)\, dt + z \int_{t_1}^{t_2} (at^z + ct^{z-1}) f(t)\, dt = 0 \tag{23}$$

Now $$z \int_{t_1}^{t_2} (at^z + ct^{z-1}) f(t)\, dt = z \int_{t_1}^{t_2} t^{z-1}(at+c) f(t)\, dt$$

$$= [t^z(at+c)f(t)]_{t_1}^{t_2} - \int_{t_1}^{t_2} t^z \frac{d}{dt}[(at+c)f(t)]\, dt \tag{24}$$

on integration by parts. Hence (23) becomes, on inserting (24),

$$\int_{t_1}^{t_2} t^{z-1} \left[(t^2 + bt + d) f(t) - t \frac{d}{dt}\{(at+c)f(t)\} \right] dt$$
$$+ [t^z(at+c)f(t)]_{t_1}^{t_2} = 0.$$

We now choose t_1, t_2, and $f(t)$ to make

$$(t^2 + bt + d) f(t) - t \frac{d}{dt}\{(at+c)f(t)\} = 0 \tag{25}$$

and $$[t^z(at+c)f(t)]_{t_1}^{t_2} = 0. \tag{26}$$

Equation (25) leads to

$$\frac{d}{dt}\{(at+c)f(t)\}/(at+c)f(t) = (t^2+bt+d)/t(at+c),$$

i.e. $$(at+c)f(t) = e^{\int (t^2+bt+d)/(at^2+ct)\, dt}$$

Now
$$\frac{t^2 + bt + d}{at^2 + ct} = A + B/t + C/(at + c)$$

where $A = 1/a, \qquad B = d/c, \qquad C = (abc - c^2 - a^2d)/ac$

Hence $\qquad (at + c)f(t) = e^{t/a} . t^{d/c}(t + c/a)^{-(c^2+da^2-abc)/a^2c}$

and we require from (26) that—
$$[e^{t/a} . t^{z+d/c}(t + c/a)^{-c/a^2-d/c+b/a}]_{t_1}^{t_2} = 0.$$

If t_2 and t_1 are chosen to satisfy this condition then
$$y(z) = \int_{t_1}^{t_2} dt\, t^{z-1+d/c} . e^{t/a}(at + c)^{\frac{b-c-a}{a} - \frac{d}{c}}$$

will satisfy the equation (21).

Example
$$y(z + 2) - 2(z + 1)y(z + 1) + (z + 1)y(z) = 0.$$

Here $\qquad a = -2, \qquad b = -2, \qquad c = 1, \qquad d = 1.$

Hence we require $[e^{-t/2} . t^{z+1} . (t - 1/2)^{1/2}]_{t_2}^{t_1} = 0.$

Thus if z lies in the part of the plane to the right of $z = -1$, we may take $t_2 = \infty$ and $t_1 = 1/2$, and a solution is
$$y(z) = \int_{\frac{1}{2}}^{\infty} dt\, t^z e^{-t/2}(t - \tfrac{1}{2})^{-\frac{1}{2}}.$$

Series Solutions of Linear Difference Equations

Expansion in Ascending or Descending Power Series of a Parameter

Under certain conditions it is possible to find solutions of the form
$$y(x) = y_0(x) + \lambda y_1(x) + \lambda^2 y_2(x) + \ldots$$

and/or
$$y(x) = y_0(x) + \frac{1}{\lambda} y_1(x) + \frac{1}{\lambda^2} y_2(x) + \ldots$$

which satisfy second order linear equations of the type
$$0 = y(x + 2) + p(x, \lambda)y(x + 1) + q(x, \lambda)y(x),$$

where λ is some parameter which occurs in (or may be conveniently inserted into) the equation.

We do not propose here to discuss in detail the classes of functions $p(x, \lambda)$ and $q(x, \lambda)$ which yield the various forms of series solutions. We propose merely to illustrate with a few examples.

Examples

1. Consider the equation

$$\Delta^2 y(x) - \lambda a^x y(x) = 0.$$

We write $$y(x) = y_0(x) + \lambda y_1(x) + \lambda^2 y_2(x) + \dots .$$

Substituting, we have,

$$\Delta^2 y_0(x) + \lambda \Delta^2 y_1(x) + \lambda^2 \Delta^2 y_2(x) + \dots = \lambda a^x y_0(x) + \lambda^2 a^x y_1(x) + \dots .$$

Equating coefficients of λ we obtain

$$\Delta^2 y_0(x) = 0.$$

Thus we may take $$y_0(x) = 1.$$

Further $$\Delta^2 y_1(x) = a^x y_0(x),$$

so that $$y_1(x) = a^x/(a-1)^2.$$

Similarly, we have, $$\Delta^2 y_2(x) = a^{2x}/(a-1)^2,$$

giving $$y_2(x) = a^{2x}/(a-1)^2(a^2-1)^2.$$

Thus we obtain—

$$y(x) = 1 + \frac{\lambda a^x}{(a-1)^2} + \frac{\lambda^2 a^{2x}}{(a-1)^2(a^2-1)^2} + \dots + \frac{\lambda^r a^{rx}}{(a-1)^2(a^2-1)^2 \dots (a^r-1)^2} + \dots .$$

This series converges for all x and λ provided $|a| \geqslant 1$. If $|a| < 1$, the series converges for such values of x as will make $|\lambda a^x| < 1$.

2. *Analogue to Bessel's equation of order n.* Consider,

$$\Delta[(x+1)(x+2) \dots (x+2n+1)\Delta y(x)] + \lambda(x+1)(x+2) \dots (x+2n+1)y(x) = 0,$$

or

$$(x+2n+2)y(x+2) - (2x+2n+3)y(x+1) + (x+1)(1+\lambda)y(x) = 0,$$

where n is an integer. We write,

$$y(x) = y_0(x) + \lambda y_1(x) + \lambda^2 y_2(x) + \ldots + \lambda^r y_r(x) + \ldots$$

as before. Then

$$\Delta[(x+1)(x+2)\ldots(x+2x+1)\Delta y_0(x)] = 0$$

so we may take $\qquad y_0(x) = 1.$

Thus we have

$$\Delta[(x+1)(x+2)\ldots(x+2n+1)\Delta y_1(x)]$$
$$= -(x+1)\ldots(x+2n+1),$$

from which it follows that we may take

$$y_1(x) = -x(x-1)/2(2n+2).$$

Proceeding in this way, the following series may be obtained—

$$y(x) = 1 + \sum_{r=1}^{\infty} (-1)^r \lambda^r \frac{x(x-1)\ldots(x-2r+1)}{2.4\ldots 2n(2n+2)\ldots(2n+2r)}$$

which is convergent for all x and n provided $|\lambda| < 1$. The similarity in form of this series with that for $J_n(x\sqrt{\lambda})$, viz.

$$J_n(x\sqrt{\lambda}) = \frac{\lambda^{n/2}x^n}{2^n n!}\left[1 - \frac{\lambda x^2}{2(2n+2)} + \frac{\lambda^2 x^4}{2.4.(2n+2)(2n+4)} - \ldots\right]$$

is immediately apparent.

3. Consider the equation—

$$(x+1)y(x+2) - (2x+1-\lambda)y(x+1) + xy(x) = 0,$$

or $$\Delta[x\Delta y(x)] + \lambda y(x+1) = 0.$$

Writing as before,

$$y(x) = y_0(x) + \lambda y_1(x) + \lambda^2 y_2(x) + \ldots$$

and inserting this series for $y(x)$ into the equation, we obtain,

$$\Delta[x\Delta y_0(x)] = 0;$$

hence we may take, $\qquad y_0(x) = 1.$

Equating coefficients of λ to zero, we obtain,

$$-y_0(x+1) - \Delta[x\Delta y_1(x)] = 0,$$

hence we may take $\Delta y_1(x) = -1,$
and $y_1(x) = -x.$

Proceeding in this way, we obtain,

$$y_r(x) = \frac{(-1)^r x(x+1)(x+2)\ldots(x+r-1)}{2^2 \cdot 3^2 \cdot 4^2 \ldots r^2}.$$

Hence,

$$y(x) = 1 - \lambda x + \lambda^2 x(x+1)/2^2 - \lambda^3 x(x+1)(x+2)/2^2 \cdot 3^2 + \ldots$$
$$= 1 + \sum_0^\infty [\lambda^{r+1} x(x+1)\ldots(x+r)/2^2 \cdot 3^2 \ldots (r+1)^2 (-1)^{r+1}]$$

is a solution of the equation

$$\Delta[x\Delta y(x)] + \lambda y(x+1) = 0.$$

EXERCISE (X)

Show that the above series is absolutely convergent for all values of x and λ.

For negative integral values of x the above series terminates, and the difference equation has in that case a polynomial solution. Furthermore as far as x is concerned, the solution has found its expression in a series of ascending factorials, instead of, as in the case of the solution of a differential equation, in a series of ascending powers of x. The difference and the similarity stand out clearly when we compare this difference equation, which, as we have seen, can be written in the form—

$$\Delta[x\Delta y(x)] + \lambda y(x+1) = 0,$$

and its solution

$$y(x) = 1 - \frac{\lambda x}{1^2} + \frac{\lambda^2 x(x+1)}{1^2 \cdot 2^2} - \frac{\lambda^3 x(x+1)(x+2)}{1^2 \cdot 2^2 \cdot 3^2} + \ldots,$$

with the differential equation

$$\frac{d}{dx}\left(x\frac{dy}{dx}\right) + \lambda y = 0,$$

and its solution

$$y = 1 - \frac{\lambda x}{1^2} + \frac{\lambda^2 x^2}{1^2 \cdot 2^2} - \frac{\lambda^3 x^3}{1^2 \cdot 2^2 \cdot 3^2} + \ldots$$

which are Bessel's equations of order zero with $\lambda\sqrt{x}$ for x, and $J_0(\lambda\sqrt{x})$ respectively. From the difference equation we have derived an analogue to the Bessel function, and we may expect it to possess corresponding analogous properties.

Solution in Factorial Series

We consider a new operator ρ defined by—

$$\rho = x\text{E},$$

i.e.
$$\rho y(x) = xy(x + 1).$$

Then
$$\rho^2 = x\text{E}(x\text{E}),$$

i.e.
$$\rho^2 y(x) = x(x + 1)y(x + 2);$$

and generally if r is a positive integer

$$\rho^r y(x) = x(x + 1) \ldots (x + r - 1)y(x + r).$$

We notice that when, but only when, $y(x) \equiv 1$ then

$$\begin{aligned}\rho &= x,\\ \rho^2 &= x(x + 1),\\ \rho^r &= x(x + 1) \ldots (x + r - 1),\end{aligned}$$

where we write $\rho . 1 = \rho$, $\rho^2 . 1 = \rho^2$, etc. Certain special types of difference equations take a particularly simple form when the operator ρ is used. Thus for example

$$x(x + 1)y(x + 2) - 5xy(x + 1) + 6y(x) = 0,$$

becomes
$$(\rho^2 - 5\rho + 6)y(x) = 0.$$

It is obvious that ρ satisfies the rules of simple algebra—

$$\rho + a = a + \rho,$$

$$a\rho = \rho a,$$

since
$$a\rho y(x) = \rho a y(x),$$

$$\rho(a + \rho) = a\rho + \rho^2,$$

if a is independent of x. Again

$$\rho^m . \rho^n = \rho^{m+n} = \rho^n . \rho^m,$$

where m and n are positive integers or zero.

For the moment we restrict ourselves to positive powers of ρ, and the rules of simple algebra apply. Thus the equation we have used in illustration, viz.

$$(\rho^2 - 5\rho + 6)y(x) = 0$$

may be written either as

$$(\rho - 3)(\rho - 2)y(x) = 0,$$

or

$$(\rho - 2)(\rho - 3)y(x) = 0.$$

It follows in fact that the two equations

$$(\rho - 2)y(x) = 0,$$

and

$$(\rho - 3)y(x) = 0,$$

will both provide possible solutions of the original equations. The equation

$$(\rho - a)y(x) = 0$$

is

$$xy(x + 1) = ay(x).$$

Its solution is therefore $\quad y(x) = Aa^x/\Gamma(x)$

where A is a unit periodic.

Theorem

If a difference equation can be written symbolically in the form

$$(\rho^r + \lambda_1\rho^{r-1} + \lambda_2\rho^{r-2} + \ldots + \lambda_r)y(x) = 0$$

where $\lambda_1, \lambda_2, \ldots, \lambda_r$ are constants then, by factorizing the polynomial in ρ into the form

$$(\rho - a_1)(\rho - a_2) \ldots (\rho - a_r)y(x) = 0,$$

the solution will be

$$y(x) = \left(\sum_{n=1}^{r} A_n a_n^x\right)/\Gamma(x)$$

where $A_1, A_2, \ldots, A_r$ are arbitrary unit periodics. Written out fully the difference equation whose solution is given above is

$$\begin{aligned} x(x+1)\ldots(x+r-1)y(x+r) \\ + \lambda_1 x(x+1)\ldots(x+r-2)y(x+r-1) + \ldots \\ + \lambda_{r-1}xy(x+1) + \lambda_r y(x) = 0. \end{aligned}$$

Now consider

$$x(x+1)y(x+2) - 2axy(x+1) + a^2y(x) = 0,$$

i.e.
$$(\rho^2 - 2a\rho + a^2)y(x) = 0,$$

or
$$(\rho - a)^2y(x) = 0.$$

Let
$$(\rho - a)y(x) = z(x)$$

then
$$(\rho - a)z(x) = 0,$$

i.e.
$$z(x) = A_1a^x/\Gamma(x)$$

as before. Hence

$$(\rho - a)y(x) = z(x) = A_1a^x/\Gamma(x)$$

or
$$xy(x+1) - ay(x) = A_1a^x/\Gamma(x)$$

or
$$\frac{\Gamma(x+1)y(x+1)}{a^x} - \frac{\Gamma(x)y(x)}{a^{x-1}} = A_1.$$

Thus
$$\Delta[\Gamma(x)y(x)/a^{x-1}] = A_1,$$

i.e.
$$\Gamma(x)y(x)/a^{x-1} = A_1x + B_1,$$

or
$$y(x) = a^x(Ax + B)/\Gamma(x),$$

where A and B are arbitrary unit periodics. Generally if

$$(\rho - a)^r y(x) = 0,$$

then
$$y(x) = \frac{a^x}{\Gamma(x)} \cdot \sum_0^{r-1} A_n x^n,$$

where $A_n (n = 0, 1, \ldots, r-1)$ are arbitrary unit periodics.

We have seen that *when the function operated upon is unity*, then—

$$\rho \,.\, 1 = x, \qquad \rho^2 \,.\, 1 = x(x+1), \qquad \rho^3 \,.\, 1 = x(x+1)(x+2), \text{ etc.}$$

or alternatively

$$x = \rho \,.\, 1, \qquad x^2 = \rho(\rho - 1) \,.\, 1, \qquad x^3 = \rho(\rho^2 - 3\rho + 1) \,.\, 1, \text{ etc.}$$

For convenience as already indicated we propose to drop the unit symbol on which the operation takes place, but it must always be remembered that it is assumed to be present.

It is clear that since ρ^n is expressible in terms of a product of n linear factors of x, any function of x of degree n will be expressible as a function of ρ of degree n operating on unity; and vice versa.

Thus if $f(x)$ is a polynomial function we can write

$$f(x) = a_0 + a_1\rho + a_2\rho^2 + a_3\rho^3 + \ldots$$
$$= a_0 + a_1x + a_2x(x+1) + a_3x(x+1)(x+2) + \ldots.$$

In fact, of course, the coefficients a_0, a_1, a_2 etc. are easily determined. Thus if we insert $x = 0$, we find $a_0 = f(0)$. If $x = -1$ we obtain

$$a_1 = f(0) - f(-1) = \Delta f(-1),$$

$x = -2$ gives $a_2 = \Delta^2 f(-2)/2!$.

Indeed, as can easily be established

$$f(x) = f(0) + \sum_1 \frac{\Delta^n f(-n)}{n!} x(x+1)\ldots(x+n-1),$$

and this will be valid for a series of unlimited number of terms provided that series is convergent.

Thus $$f(x) = f(0) + \sum_1^n \frac{\Delta^n f(-n)}{n!} \rho^n,$$

where the right-hand side is assumed to be operating on unity.

EXERCISE (XI)

Obtain this expression directly by using the symbolical relation

$$E^x = (1 - \Delta E^{-1})^x,$$

expanding the right-hand side, operating with both sides on $f(a)$ and then writing $a = 0$.

Here we are primarily concerned with the operator ρ and the circumstances in which it may be used to special advantage. To illustrate this consider the equation

$$xy(x+1) - 2y(x) = x^3 + 2x^2 - 2.$$

The left-hand side may be written $(\rho - 2)y(x)$ and therefore the complementary function is, as we have seen, $A2^x/\Gamma(x)$, where A is a unit periodic.

The right-hand side may be written

$$x^3 + 2x^2 - 2 = \rho(\rho^2 - 3\rho + 1) + 2\rho(\rho - 1) - 2$$
$$= \rho^3 - \rho^2 - \rho - 2 = (\rho - 2)(\rho^2 + \rho + 1)$$

where the ρ's are regarded as operating on unity. Thus the equation is

$$(\rho - 2)y(x) = (\rho - 2)(\rho^2 + \rho + 1) \,.\, 1.$$

Accordingly a particular solution is clearly

$$y(x) = (\rho^2 + \rho + 1) \,.\, 1 = x(x+1) + x + 1 = (x+1)^2$$

Thus finally $$y(x) = A2^x/\Gamma(x) + (x+1)^2.$$

It is clear that no difficulty arises in this case in determining the particular integral because the factor on the left $\rho - 2$ is also present on the right. Suppose, however, that this is not the case, and consider

$$(\rho - a)y(x) = f(\rho),$$

it being assumed that the function of x on the right has been converted into a function of ρ operating on unity. The particular solution may now be written symbolically as

$$f(\rho)/(\rho - a) \,.\, 1$$

provided this expression can be consistently interpreted, and that the result of the operation is sensible.

Now $f(\rho)/(\rho - a)$ may be expanded either in ascending or descending powers of ρ. Thus

1. $$f(\rho)/(\rho - a) = a_0\rho^m + a_1\rho^{m+1} + a_2\rho^{m+2} + \ldots$$
$$= a_0x(x+1)\ldots(x+m-1) + a_1x(x+1)\ldots(x+m)$$
$$+ a_2x(x+1)\ldots(x+m+1) + \ldots.$$

Such a formal series may not be convergent since the ratio test demands that

$$|a_{r+1}(x + m + r)/a_r| < 1$$

for all values of r beyond a certain value, however large. It will, however, certainly provide a finite set of terms *if x is a negative integer.*

2. $$f(\rho)/(\rho - a) = \phi(\rho) + a_1\rho^{-1} + a_2\rho^{-2} + \ldots$$

where $\phi(\rho)$ is a polynomial in ρ of finite degree. So far we cannot carry through these operations on unity since we have not yet interpreted $\rho^{-1}, \rho^{-2}, \ldots.$

To do so, and to justify the expansion of a function of ρ in negative powers we must define ρ^{-1} in such a manner as to preserve the law of indices for negative powers. Accordingly we define ρ^{-1} so that

$$\rho \,.\, \rho^{-1} = 1,$$

or generally $$\rho[\rho^{-1}y(x)] = y(x).$$

This requires that

$$x\mathrm{E}[\rho^{-1}y(x)] = y(x),$$

i.e.

$$\mathrm{E}[\rho^{-1}y(x)] = y(x)/x,$$

i.e.

$$\rho^{-1}y(x) = \mathrm{E}^{-1}[y(x)/x] = (x-1)^{-1}y(x-1).$$

This provides a definite and unique meaning to the operator of ρ^{-1}, and demands that

$$\rho^{-1}\,.\,1 = \frac{1}{x-1}.$$

Now let us examine what relation exists between $\rho\rho^{-1}$ and $\rho^{-1}\rho$. These should be identical if the index law is to be valid in this case. Now from what we have seen

$$\rho^{-1}\rho y(x) = \rho^{-1}[xy(x+1)] = \mathrm{E}^{-1}\left[\frac{xy(x+1)}{x}\right] = y(x).$$

It follows that $\quad \rho^{-1}\rho y(x) = y(x) = \rho\rho^{-1}y(x)$

and the operators ρ and ρ^{-1} are commutative.

We are now in a position to determine not merely the meaning of $\rho^{-1}\,.\,1$ which we found to be $1/(x-1)$ but $\rho^{-r}\,.\,1$ where r is a positive integer. For

$$\begin{aligned}\rho^{-2}y(x) &= \rho^{-1}\rho^{-1}y(x)\\ &= \rho^{-1}(x-1)^{-1}y(x-1)\\ &= (x-1)^{-1}(x-2)^{-1}y(x-2).\end{aligned}$$

Generally

$$\begin{aligned}\rho^{-r}y(x) &= (x-r)^{-1}(x-r+1)^{-1}(x-r+2)^{-1}\ldots(x-1)^{-1}y(x-r),\\ &= y(x-r)/(x-r)(x-r+1)\ldots(x-1).\end{aligned}$$

Accordingly

$$\rho^{-r} = \rho^{-r}\,.\,1 = 1/(x-r)(x-r+1)\ldots(x-1).$$

Returning now to the second mode of treatment of the particular solution of

$$(\rho - a)y(x) = f(\rho)$$

giving $\quad y(x) = f(\rho)/(\rho - a) = \phi(\rho) + a_1\rho^{-1} + a_2\rho^{-2} + \ldots,$

we know that $\phi(\rho)\,.\,1$ is easily interpreted as before, while the terms

$$(a_1\rho^{-1} + a_2\rho^{-2} + \ldots)\,.\,1 = a_1/(x-1) + a_2/(x-1)(x-2) + \ldots$$

give a series that is in general convergent provided x is not a positive integer.

The method of determining the complementary function and the particular solution may now clearly be applied, apart from certain apparently exceptional cases, to the general difference equation

$$\phi(\rho)y(x) = f(\rho)$$

where $\phi(\rho)$ is a polynomial of finite degree in ρ.

It is, however, true that this is a restricted type of equation. The illustration with which we commenced this section, viz.

$$(x + 1)y(x + 2) - (2x + 1 - \lambda)y(x + 1) + xy(x) = 0$$

does not in fact belong to this category. It cannot be written directly in the form $f(\rho)y(x) = 0$. Nevertheless, as we have seen, it does possess a solution which is expressible as a power series in ρ operating on 1.

To enlarge the field of operations therefore we return to our definition

$$\rho = x\mathrm{E} = x(1 + \Delta) = x + x\Delta.$$

We propose to regard $x\Delta$ as an associated operator π, so that

$$\rho - \pi = x.$$

The operator π satisfies the ordinary rules of algebra. Thus

$$(\pi + a)y(x) = \pi y(x) + ay(x) = (a + \pi)y(x)$$

$$a\pi y(x) = \pi a y(x)$$

so that

$$\pi + a = a + \pi$$

$$\pi a = a\pi$$

provided here a is a constant. Moreover the ordinary law of indices applies, viz.

$$\pi^r \,.\, \pi^s y(x) = \pi^{r+s} y(x)$$

where r and s are positive integers. Also the distributive law applies here, for—

$$(\pi + a)(\pi + b) \equiv (\pi + b)(\pi + a)$$

provided a and b are constants.

We now turn to an examination of the interaction of ρ and π. In the first place we have seen that in general

$$\rho - \pi = x.$$

Accordingly $\quad \rho y(x) = xy(x+1)$

$$\pi y(x) = x\Delta y(x) = xy(x+1) - xy(x).$$

Hence $\quad \pi\rho y(x) = x\Delta[xy(x+1)]$

$$= x(x+1)y(x+2) - x^2y(x+1)$$

by operating on both sides of the first of these with π. Again, by operating on the second with ρ we have

$$\rho\pi y(x) = x\mathrm{E}[xy(x+1) - xy(x)]$$

$$= x(x+1)y(x+2) - x(x+1)y(x+1).$$

Hence $\quad (\pi\rho - \rho\pi)y(x) = xy(x+1) = x\mathrm{E}y(x) = \rho y(x).$

Accordingly, $\quad \pi\rho - \rho\pi \equiv \rho,$

and therefore ρ and π are *not* commutative. We must take particular care to take account of their relative order. If we wish to change the order of $\pi\rho$ as an operator we must in fact add ρ to $\rho\pi$.

EXERCISES (XII)

1. Show that

(i) $\pi^2y(x) = x(x+1)\Delta^2y(x) + x\Delta y(x).$

(ii) $\pi^3y(x) = x(x+1)(x+2)\Delta^3y(x) + 3x(x+1)\Delta^2y(x) + x\Delta y(x).$

(iii) $x(x+1)\Delta^2y(x) = \pi(\pi-1)y(x).$

(iv) $x(x+1)(x+2)\Delta^3y(x) = \pi(\pi-1)(\pi-2)y(x).$

2. Show by induction that

$$x(x+1)\ldots(x+n-1)\Delta^ny(x) = \pi(\pi-1)\ldots(\pi-n+1)y(x).$$

3. Show that

$$(\pi-a)(\pi-b) \equiv (\pi-b)(\pi-a),$$

$$(\rho-a)(\rho-b) \equiv (\rho-b)(\rho-a),$$

but $\quad (\pi-a)(\rho-b) \not\equiv (\rho-b)(\pi-a).$

4. If $\quad (\pi-a)(\rho-b)y(x) = (\rho-a)(\pi-b)y(x)$

show that $\quad y(x) = A(a-b)^x.$

Some Properties of the Operators ρ and π

We have seen that

$$\pi\rho - \rho\pi = \rho \qquad \text{or} \qquad \rho\pi = (\pi-1)\rho.$$

Accordingly

$$\rho^2\pi = \rho[(\pi-1)\rho] = (\rho\pi-\rho)\rho = [(\pi-1)\rho - \rho]\rho = (\pi-2)\rho^2.$$

By induction we can easily show that

$$\rho^n\pi = (\pi-n)\rho^n,$$

where n is an integer. For, assuming this true for any particular n, it follows that

$$\begin{aligned}\rho^{n+1}\pi &= \rho(\pi - n)\rho^n = (\rho\pi - n\rho)\rho^n \\ &= [(\pi - 1)\rho - n\rho]\rho^n = [\pi - (n+1)]\rho^{n+1}.\end{aligned}$$

Thus if the proposition is true for n, it is true for $n + 1$; but it is true for $n = 1$ and $n = 2$. It is therefore true for $n = 3, 4, \ldots$.

We are accordingly entitled to bring π from the right-hand side to the left-hand side of ρ^n provided we diminish π at the same time by n. And, conversely, we may bring π from the left to the right of ρ^n if we change π to $\pi + n$.

Now consider $\rho\pi^2$. Since

$$\begin{aligned}\rho\pi &= (\pi - 1)\rho \\ \rho\pi^2 &= (\pi - 1)\rho\pi = (\pi - 1)\,.\,(\pi - 1)\rho = (\pi - 1)^2\rho;\end{aligned}$$

and generally if m is an integer

$$\rho\pi^m = (\pi - 1)^m\rho.$$

In the same way since $\rho^n\pi = (\pi - n)\rho^n$,

then
$$\begin{aligned}\rho^n\pi^2 &= (\pi - n)\rho^n\pi = (\pi - n)(\pi - n)\rho^n \\ &= (\pi - n)^2\rho^n.\end{aligned}$$

And generally if m and n are any integers

$$\rho^n\pi^m = (\pi - n)^m\rho^n$$

or, alternatively, $\pi^m\rho^n = \rho^n(\pi + n)^m$.

The simple way to regard this is to say that if π passes to the positive side of ρ^n then π^m changes to $(\pi + n)^m$, and if to the negative side, to $(\pi - n)^m$.

More generally, if $f(\pi)$ is a polynomial function of π, since this rule applies to each term in the expression, then

$$f(\pi)\rho^n = \rho^n f(\pi + n)$$

and

$$\rho^n f(\pi) = f(\pi - n)\rho^n$$

where n is an integer.

Finally, therefore, we have the relations

$$\rho - \pi = x$$

$$f(\pi)\rho^n = \rho^n f(\pi + n).$$

The first of these enables us to represent any function of x as a function of ρ and π, and the second will enable us to bring every term in such a function into a standard form.

A case of great importance for further development arises when the function operated on is 1. As we have already seen

$$\rho \,.\, 1 = x$$
$$\rho^n \,.\, 1 = x(x+1) \,.\,.\,.\, (x+n-1).$$

Here we have

$$\begin{aligned} f(\pi)\rho^n \,.\, 1 &= \rho^n f(\pi + n) \,.\, 1 = \rho^n f(n) \,.\, 1 \\ &= f(n) \,.\, x(x+1) \,.\,.\,.\, (x+n-1) \end{aligned}$$

since, when $f(\pi + n)$ is expanded in powers of π, every term operating on unity is zero except that involving $f(n)$.

We are now in a position to see how these properties of ρ and π may be used to determine the solution of a linear difference equation whose coefficients are polynomial functions of x of finite degree. Consider the linear difference equation of the rth order

$$y(x+r) + p_1(x)y(x+r-1) + \,.\,.\,.\, + p_r(x)y(x) = R(x)$$

or, say,
$$\phi(x, \mathrm{E})y = R(x).$$

It is not difficult to see—and we shall illustrate the procedure in particular cases—that this may be written in the form—

$$\psi(x, x\mathrm{E}) = R(x)$$

or
$$\psi(x, \rho)y(x) = R(x).$$

Since $\psi(x, \rho)$ operates on a function of x we may replace x by $\rho - \pi$. On the other hand, since $R(x)$ operates on 1, we may express $R(x)$ as $R_1(\rho)$. Moreover, since $y(x)$ can be regarded as operating on 1, we shall seek for solutions of the form, if such exist,

$$\begin{aligned} y(x) &= a_0 + a_1\rho + a_2\rho^2 + \,.\,.\,. \\ &= a_0 + a_1x + a_2x(x+1) + a_3x(x+1)(x+2) + \,.\,.\,.\,. \end{aligned}$$

Accordingly the original equation may be recast in the form

$$\psi[(\rho - \pi), \rho] \,.\, (a_0 + a_1\rho + a_2\rho^2 + \,.\,.\,.\,) = R_1(\rho).$$

Using the relation that for every integral value of n,

$$f(\pi)\rho^n \,.\, 1 = \rho^n f(n),$$

the left-hand side may finally be written as a power series in ρ. By equating the coefficients to zero we determine $a_0, a_1, a_2 \ldots$ and so the solution to the complementary equation is found if it exists in this form. On the other hand, by equating the coefficients of powers of ρ to the corresponding coefficients of powers of ρ on the right, the particular integral is found. In this way the solution, if it exists, is determined. Before examining in what circumstances this procedure will be successful, we shall illustrate with examples in which it is valid.

Consider again, the equation

$$(x+1)y(x+2) - (2x+1-\lambda)y(x+1) + xy(x) = 0.$$

Here $$[(x+1)\mathrm{E}^2 - (2x+1-\lambda)\mathrm{E} + x]y(x) = 0,$$

or $$[x(x+1)\mathrm{E}^2 - (2x+1-\lambda)x\mathrm{E} + x^2]y(x) = 0,$$

or $$[\rho^2 - (2\rho - 2\pi + 1 - \lambda)\rho + (\rho - \pi)(\rho - \pi)]y(x) = 0.$$

The operator is

$$\rho^2 - 2\rho^2 + 2\pi\rho - (1-\lambda)\rho + \rho^2 - \rho\pi - \pi\rho + \pi^2$$
$$= \pi\rho - (\pi - 1)\rho - (1-\lambda)\rho + \pi^2 = \lambda\rho + \pi^2.$$

Thus the equation reduces to

$$(\lambda\rho + \pi^2)y(x) = 0.$$

Write $y(x)$ tentatively in the form

$$y(x) = a_0 + a_1\rho + a_2\rho^2 + \ldots + a_n\rho^n + \ldots$$

and insert in the equation, then

$$\lambda a_0\rho + \lambda a_1\rho^2 + \lambda a_2\rho^3 + \ldots + \lambda a_{n-1}\rho^n + \ldots + \pi^2 a_0 + a_1\pi^2\rho$$
$$+ a_2\pi^2\rho^2 + a_3\pi^2\rho^3 + \ldots + a_n\pi^2\rho^n + \ldots = 0,$$

every term presumed to be operating on unity. Now

$$\pi^2 a_0 \,.\, 1 = 0,$$
$$\pi^2\rho \,.\, 1 = \rho(\pi+2)^2 \,.\, 1 = 2^2\rho, \ldots, \pi^2\rho^n \,.\, 1 = n^2\rho^n \ldots.$$

Accordingly $$2^2 a_1 = -\lambda a_0,$$
$$3^2 a_2 = -\lambda a_1,$$
$$\ldots\ldots\ldots\ldots$$
$$n^2 a_n = -\lambda a_{n-1},$$

on equating powers of ρ to zero. Hence a_0 is arbitrary,

$$a_1 = -a_0\lambda/2^2,\ a_2 = a_0\lambda^2/2^2 \,.\, 3^2 \ldots,$$

and consequently

$$\begin{aligned} y(x) &= a_0[1 - \lambda\rho/2^2 + \lambda^2\rho^2/2^2 3^2 - \lambda^3\rho^3/2^2 \,.\, 3^2 \,.\, 4^2 + \ldots] \\ &= a_0[1 - \lambda x/2^2 + \lambda^2 x(x+1)/2^2 \,.\, 3^2 \\ &\qquad - \lambda^3 x(x+1)(x+2)/2^2 \,.\, 3^2 \,.\, 4^2 + \ldots] \end{aligned}$$

which is the same expansion as we have already derived.

To illustrate the derivation of the particular solution, suppose that $\lambda = -2$ in the above and that we have the equation

$$(x+1)y(x+2) - (2x+3)y(x+1) + xy(x) = x^2.$$

Proceeding as before, we obtain

$$[x(x+1)E^2 - (2x+3)xE + x^2]y(x) = x^3$$

or

$$\begin{aligned} [\rho^2 - (2\rho - 2\pi + 3)\rho + (\rho - \pi)(\rho - \pi)]y(x) & \\ = (\rho - \pi)(\rho - \pi)(\rho - \pi). & \end{aligned}$$

The operator on the left reduces to $(-2\rho + \pi^2)$. On the right

$$\begin{aligned} (\rho - \pi)(\rho - \pi)(\rho - \pi) \,.\, 1 &= (\rho^3 - \rho\pi\rho - \pi\rho^2 + \pi^2\rho) \,.\, 1 \\ &= [\rho^3 - \rho^2(\pi+1) - \rho^2(\pi+2) + \rho(\pi+1)^2] \,.\, 1 \\ &= \rho^3 - 3\rho^2 + \rho. \end{aligned}$$

Now assume tentatively for the particular solution

$$y(x) = A_0 + A_1\rho + A_2\rho^2 + \ldots.$$

Hence, as before

$$\begin{aligned} \pi^2 A_0 + (A_1\pi^2 - 2A_0)\rho + (A_2\pi^2 - 2A_1)\rho^2 & \\ + (A_3\pi^2 - 2A_2)\rho^3 + \ldots = \rho - 3\rho^2 + \rho^3, & \end{aligned}$$

both sides presumed to be operating on unity. Accordingly, $\pi^2 A_0 = 0$, and A_0 is undetermined.

$$\begin{aligned} (A_1\pi^2 - 2A_0)\rho &= \rho, &\quad \text{i.e.}\quad A_1 - 2A_0 &= 1 \\ (A_2\pi^2 - 2A_1)\rho^2 &= -3\rho^2, &\quad \text{i.e.}\quad 4A_2 - 2A_1 &= -3 \\ (A_3\pi^2 - 2A_2)\rho^3 &= \rho^3, &\quad \text{i.e.}\quad 9A_3 - 2A_2 &= 1 \\ (A_4\pi^2 - 2A_3)\rho^4 &= 0, &\quad \text{i.e.}\quad 16A_4 - 2A_3 &= 0 \end{aligned}$$

. . . .

Thus $$\begin{aligned} A_1 &= 2A_0 + 1 \\ A_2 &= (4A_0 - 1)/4 \\ A_3 &= (4A_0 + 1)/18 \\ A_4 &= (4A_0 + 1)/144. \end{aligned}$$

It is clear that if $A_3 = 0$ then all succeeding A's are zero. Accordingly we choose $A_0 = -\frac{1}{4}$, then $A_1 = \frac{1}{2}$, and $A_2 = -\frac{1}{2}$, so that the particular integral is

$$-\frac{1}{4} + \frac{1}{2}\rho - \frac{1}{2}\rho^2 = -\frac{1}{4} + \frac{x}{2} - \frac{x(x+1)}{2} = -\frac{x^2}{2} - \frac{1}{4}.$$

Verification

$$\begin{aligned} &(x+1)y(x+2) - (2x+3)y(x+1) + xy(x) \\ &\quad = (x+1)\left[-\frac{(x+2)^2}{2} - \frac{1}{4}\right] - (2x+3)\left[-\frac{(x+1)^2}{2} - \frac{1}{4}\right] \\ &\qquad + x\left[-\frac{x^2}{2} - \frac{1}{4}\right] \\ &\quad = \tfrac{1}{4}[-(x+1)(2x^2+8x+9) + (2x+3)(2x^2+4x+3) \\ &\qquad - x(2x^2+1)] \\ &\quad = \tfrac{1}{4}[4x^2] = x^2. \end{aligned}$$

Consider more generally an equation of the form

$$[\rho + f(\pi)]y(x) = 0.$$

Let $$y(x) = a_0 + a_1\rho + a_2\rho^2 + \ldots.$$

Hence—

$$\begin{aligned} a_0\rho + a_1\rho^2 + a_2\rho^3 + \ldots + a_{n-1}\rho^n + \ldots + a_0 f(0) + a_1 f(1)\rho \\ + a_2 f(2)\rho^2 + a_3 f(3)\rho^3 + \ldots + a_n f(n)\rho^n + \ldots = 0. \end{aligned}$$

Thus we require—

$$\begin{aligned} a_0 f(0) &= 0, \\ a_1 f(1) + a_0 &= 0, \\ a_2 f(2) + a_1 &= 0, \\ &\ldots\ldots\ldots \\ a_n f(n) + a_{n-1} &= 0, \\ &\ldots\ldots\ldots \end{aligned}$$

If $f(n)$ is not zero for any integral value of n then

$$a_0 = a_1 = a_2 \ldots = 0,$$

and there is no solution as a power series in ρ. If $f(n)$ is zero where $n = 0$, or 1, or 2, . . ., or r, then the series in ρ commences with ρ^{r+1}, a formal solution can then be obtained and the question of convergence will arise. Suppose, for example, that $f(1) = 0$, then

$$y(x) = a_1\left[-\frac{\rho^2}{f(2)} + \frac{\rho^3}{f(2)f(3)} - \frac{\rho^4}{f(2)f(3)f(4)} + \ldots\right]$$

$$= A\left[\frac{x(x+1)}{f(2)} - \frac{x(x+1)(x+2)}{f(2)f(3)} + \frac{x(x+1)(x+2)(x+3)}{f(2)f(3)f(4)} \ldots\right].$$

The ratio test for convergence then demands that $\dfrac{x+n}{f(n+1)} < 1$ for $n > N$, and this will depend on the degree of $f(n+1)$ in n. Clearly if $f(\pi)$ is of higher degree than the first a convergent solution is readily obtained.

Example

$$x(x+1)y(x+2) - 2x^2y(x+1) + (x^2-1)y(x) = 0.$$

The operator is

$$\begin{aligned}\rho^2 - 2x\rho + x^2 - 1 &= \rho^2 - 2(\rho-\pi)\rho + (\rho-\pi)(\rho-\pi) - 1\\ &= 2\pi\rho - \pi\rho - \rho\pi + \pi^2 - 1\\ &= \pi\rho - (\pi-1)\rho + \pi^2 - 1\\ &= \rho + \pi^2 - 1.\end{aligned}$$

Hence we require

$$\begin{aligned}&(\rho + \pi^2 - 1)(a_0 + a_1\rho + a_2\rho^2 + \ldots)\\ &\quad = a_0\rho + a_1\rho^2 + a_2\rho^3 + \ldots - a_0 - a_1\rho - a_2\rho^2 - a_3\rho^3 \ldots\\ &\qquad\qquad + a_1\rho + a_2 2^2\rho^2 + a_3 3^2\rho^3 + \ldots\\ &\quad = -a_0 + a_0\rho + (a_1 - a_2 + 2^2a_2)\rho^2 + (a_2 - a_3 + 3^2a_3)\rho^3 + \ldots.\end{aligned}$$

Hence

$$\begin{aligned}a_0 &= 0,\\ a_2 &= -a_1/(2^2-1) = -a_1/1\,.\,3,\\ a_3 &= -a_2/2\,.\,4 = a_1/1\,.\,2\,.\,3\,.\,4,\\ a_4 &= -a_3/3\,.\,5 = -a_1/1\,.\,2\,.\,3^2\,.\,4\,.\,5,\\ a_5 &= -a_4/4\,.\,6 = a_1/1\,.\,2\,.\,3^2\,.\,4^2\,.\,5\,.\,6.\end{aligned}$$

$$\text{Thus } y(x) = A\left[\frac{x(x+1)}{1.3} - \frac{x(x+1)(x+2)}{1.2.3.4} + \frac{x(x+1)(x+2)(x+3)}{1.2.3^2.4.5} - \ldots\right].$$

This series is convergent since the ratio of the $(n+1)$th to the nth term is $-\dfrac{x+n}{(n+1)(n+3)}$ which rapidly becomes less than unity for any assigned value of x.

EXERCISE (XIII)

Show that it is not possible to find a particular solution of the equation—

$$x(x+1)y(x+1) - 2x^2y(x+1) + (x^2-1)y(x) = 1 + ax$$

as a series in ascending powers of ρ unless $a = -1$.

Consider now the case where the operator in the difference equation is a polynomial function of π only so that it may be written

$$(\pi - a_1)(\pi - a_2) \ldots (\pi - a_n)y(x) = 0.$$

The elementary difference equation

$$(\pi - a_r)y(x) = 0,$$

i.e. $$x\Delta y(x) - a_r y(x) = 0$$

or $$xy(x+1) = (x + a_r)y(x)$$

has as its solution $$y(x) = A_r\Gamma(x + a_r)/\Gamma(x)$$

If a_r is a positive integer then of course this reduces to a product of a_r linear factors.

In general, however, the solution of the full equation is

$$y(x) = \frac{1}{\Gamma(x)}\sum_{r=1}^{r=n} A_r\Gamma(x + a_r),$$

where A_r is a unit periodic.

EXERCISE (XIV)

Examine the nature of the solution when a_r is a negative integer.

When the operator is a function of ρ only, the solution, as we have seen for

$$(\rho - a_1)(\rho - a_2) \ldots (\rho - a_n)y(x) = 0,$$

is
$$y(x) = \frac{1}{\Gamma(x)} \sum_{1}^{n} a_r{}^x$$

and since $\Gamma(x)$, when x is large, behaves asymptotically as $x^x\,\mathrm{e}^{-x}$, we can say that the solution of the difference equation in this case behaves as

$$\mathrm{e}^x \sum_{1}^{n} (a_r/x)^x = \sum_{1}^{n} (\mathrm{e}\, a_r/x)^x$$

which diminishes to zero as x increases without limit.

It is possible frequently to introduce considerable simplification into the above process. For example, consider the comparatively simple form—

$$(\rho - 2\pi)y(x) = 0.$$

Assuming as usual an expansion for $y(x)$ of the form

$$y(x) = a_0 + a_1\rho + a_2\rho^2 + \ldots,$$

it is easily found that

$$2a_1 = a_0, \qquad 2\,.\,2a_2 = a_1, \qquad 2\,.\,3a_3 = a_2, \;\ldots$$

leading to a solution

$$y(x) = a_0\left[1 + \frac{x}{2\,.\,1!} + \frac{x(x+1)}{2^2\,.\,2!} + \frac{x(x+1)(x+2)}{2^3\,.\,3!} + \ldots\right],$$

an expression which is convergent for all values of x. Now the original equation is in reality simply

$$y(x+1) = 2y(x).$$

Its solution is therefore $A\,.\,2^x$ where A is a unit periodic, and it is easily verified that the above expansion is in fact 2^x. This suggests that had we sought a solution of the form

$$y(x) = a^x[a_0 + a_1\rho + a_2\rho^2 \ldots]$$

the whole process might have been considerably simplified provided a could be suitably chosen. Consider for example

$$x(x+1)y(x+2) - 6x(x+1)y(x+1) + 9(x^2-4)y(x) = 0.$$

We seek a solution of the form

$$y(x) = \lambda^x[a_0 + a_1\rho + a_2\rho^2 + \ldots].$$

Write $y(x) = \lambda^x z(x)$, and the equation becomes

$$\lambda^2 x(x+1)z(x+2) - 6\lambda x(x+1)z(x+1) + 9(x^2-4)z(x) = 0,$$

or, converting the operator into a function of ρ and π,

$$[(\lambda-3)^2\rho^2 - 3\{2\lambda - 3 + (6-2\lambda)\pi\}\rho + 9\pi^2 - 36]z(x) = 0.$$

Assuming $$z(x) = a_0 + a_1\rho + a_2\rho^2 + \ldots$$

and equating the coefficient of ρ^{n+2} to zero, we obtain

$$(\lambda-3)^2 a_n - 3[2\lambda - 3 + 2(3-\lambda)(n+1)]a_{n+1} + [9(n+2)^2 - 36]a_{n+2} = 0.$$

This equation reduces to one of the first order if $\lambda = 3$, in which case

$$-9a_{n+1} + [9(n+2)^2 - 36]a_{n+2} = 0,$$

or $$a_{n+1} = a_n/(n+3)(n-1)$$

giving $$a_n = A/(n+2)!(n-2)!$$

Thus $$y(x) = A3^x \sum_{n=2}^{\infty} \frac{x(x+1)\ldots(x+n-1)}{(n+2)!(n-2)!},$$

which is a rapidly convergent series. Had we not proceeded in this way but sought a direct solution

$$y(x) = A_0 + A_1\rho + A_2\rho^2 + \ldots,$$

the original equation expressed in terms of ρ and π would have given

$$[4\rho^2 - 3(4\pi - 1)\rho + 9\pi^2 - 36]y(x) = 0$$

with the relation between the coefficients

$$4A_n - 3(4n+3)A_{n+1} + 9(n+4)nA_{n+2} = 0.$$

The determination of A_n in this case would be as difficult as finding the solution of the original equation.

Solution in Inverse Factorials

Since the expansion in powers of ρ does not necessarily begin with a_0, and proceed in steps of successive powers of ρ the more general expression for $y(x)$ might be written in the form

$$y(x) = \lambda^x[a_1\rho^{m_1} + a_2\rho^{m_2} + a_3\rho^{m_3} + \ldots],$$

where $m_1, m_2, m_3, \ldots$ are integers and $m_1 < m_2 < m_3 \ldots$. We have given no interpretation to cases where any of the m's are fractional, but they could be negative integers. We need not in fact retain the restriction that $m_1 < m_2 < m_3 \ldots$ for, on occasion, it may be desirable to expand in descending powers of ρ; but since these will give rise to terms of the type

$$\frac{1}{(x-1)(x-2)\ldots(x-r)}$$

the expansion will not in general be valid for integral values of x. To meet such cases we shall have recourse to an alternative method of solution through yet another pair of operators. We define

$$\rho_1 = x\mathrm{E}^{-1},$$

and $$\pi_1 = \rho_1 - x = x(\mathrm{E}^{-1} - 1) = -x\mathrm{E}^{-1}\Delta,$$

so that in this case also $\rho_1 - \pi_1 = x$.

Moreover $$\rho_1^2 = x\mathrm{E}^{-1}(x\mathrm{E}^{-1}) = x(x-1)\mathrm{E}^{-2},$$

and $$\rho_1^r = x(x-1)\ldots(x-r+1)\mathrm{E}^{-r}.$$

Accordingly $$\rho_1 y(x) = xy(x-1);$$

$$\rho_1^r y(x) = x(x-1)\ldots(x-r+1)y(x-r).$$

With the previous convention where, when no function is present to be operated on, it is assumed that that function is unity, we have

$$\rho_1^r . 1 = x(x-1)\ldots(x-r+1).$$

For negative powers of ρ_1 we define ρ_1^{-1} by the relation

$$\rho_1\rho_1^{-1}y(x) = y(x),$$

i.e. $$x\mathrm{E}^{-1}[\rho_1^{-1}y(x)] = y(x),$$

i.e. $$\rho_1^{-1}y(x) = y(x+1)/(x+1).$$

On this basis

$$\rho_1^{-1}\rho_1 y(x) = \rho_1^{-1}[xy(x-1)] = (x+1)y(x)/(x+1) = y(x).$$

Thus again ρ_1 and ρ_1^{-1} are commutative. We can now evaluate ρ_1^{-r}, for

$$\rho_1^{-2}y(x) = \rho_1^{-1}[\rho_1^{-1}y(x)] = \rho_1^{-1}[y(x+1)/(x+1)]$$
$$= y(x+2)/(x+1)(x+2),$$

and generally for r, an integer

$$\rho_1^{-r}y(x) = y(x+r)/(x+1)(x+2)\ldots(x+r).$$

When $\quad y(x) \equiv 1$

$$\rho_1^{-r} = 1/(x+1)(x+2)\ldots(x+r),$$

as compared with

$$\rho_1^{r} = x(x-1)\ldots(x-r+1).$$

Notice also for comparison that

$$\rho^{-r} = 1/(x-1)(x-2)\ldots(x-r)$$

and

$$\rho^{r} = x(x+1)\ldots(x+r-1).$$

We turn now to examine the relation between $\rho_1\pi_1$ and $\pi_1\rho_1$. From our definitions

$$\begin{aligned}\rho_1\pi_1 y(x) &= x\mathrm{E}^{-1}[-x\mathrm{E}^{-1}\Delta]y(x)\\ &= x\mathrm{E}^{-1}[-x+x\mathrm{E}^{-1}]y(x)\\ &= \rho_1[-x+\rho_1]y(x).\end{aligned}$$

Also

$$\begin{aligned}\pi_1\rho_1 y(x) &= -x\mathrm{E}^{-1}\Delta(x\mathrm{E}^{-1})y(x)\\ &= -\rho_1(\mathrm{E}-1)[x\mathrm{E}^{-1}y(x)]\\ &= -\rho_1[x+1-x\mathrm{E}^{-1}]y(x)\\ &= \rho_1[-x-1+\rho_1]y(x)\\ &= \rho_1\pi_1 y(x) - \rho_1 y(x).\end{aligned}$$

Hence

$$\rho_1\pi_1 - \pi_1\rho_1 = \rho_1$$

when operating on $y(x)$. Thus

$$\pi_1\rho_1 y(x) = \rho_1(\pi_1 - 1)y(x).$$

Following the same procedure as before, we can easily establish that

$$\pi_1^2\rho_1 y(x) = \rho_1(\pi_1-1)^2 y(x),$$
$$\pi_1\rho_1^2 y(x) = \rho_1^2(\pi_1-2)y(x);$$

and more generally,

$$\pi_1^m\rho_1^n y(x) = \rho_1^n(\pi_1-n)^m y(x).$$

Finally as with ρ and π

$$f(\pi_1)\rho_1^n y(x) = \rho_1^n f(\pi_1 - n)y(x).$$

In the case of ρ_1 and π_1 a forward jump of $f(\pi_1)$ over ρ_1^n reduced π_1 by n instead of increasing it by this amount as in the case of ρ and π. On the other hand, the contrast is not so severe as it appears at first sight. If we are concerned, for example, with expressing the solution of the linear difference equation in the form

$$\begin{aligned} y(x) &= a_0 + a_1 x + a_2 x(x+1) + a_3 x(x+1)(x+2) + \ldots \\ &= a_0 + a_1\rho + a_2\rho^2 + a_3\rho^3 + \ldots, \end{aligned}$$

or, where this does not provide a convergent expansion, in the form

$$\begin{aligned} y(x) &= a_0 + a_1/(x+1) + a_2/(x+1)(x+2) \\ &\qquad + a_3/(x+1)(x+2)(x+3) + \ldots \\ &= a_0 + a_1\rho_1^{-1} + a_2\rho_1^{-2} + a_3\rho_1^{-3} \ldots \end{aligned}$$

then we require the "shift rule" for $f(\rho)$ as an expansion in *ascending* powers of ρ, or for $f(\rho_1)$ as an expansion in *descending* powers of ρ_1. For these cases the rule is precisely the same, viz.

$$f(\pi) \,.\, \rho^n = \rho^n f(\pi + n)$$

$$f(\pi_1)\frac{1}{\rho_1^n} = \frac{1}{\rho_1^n} f(\pi_1 + n).$$

Example

$$y(x+2) = (x+1)(x+2)\Delta y(x)$$

or

$$\frac{y(x+1)}{x+1} = x\Delta y(x-1) = x\mathrm{E}^{-1}\Delta y(x).$$

Here then $\qquad (\rho_1^{-1} - \pi_1)y(x) = 0.$

Let $\quad y(x) = a_0 + a_1\rho_1^{-1} + a_2\rho_1^{-2} + \ldots + a_r\rho_1^{-r} + \ldots,$

then on inserting in the above equation, we get

$$\begin{aligned} &a_0\rho_1^{-1} + a_1\rho_1^{-2} + a_2\rho_1^{-3} + \ldots + a_{r-1}\rho_1^{-r} + \ldots \\ &\qquad + a_1 \,.\, 1 \,.\, \rho_1^{-1} + a_2 \,.\, 2 \,.\, \rho_1^{-2} + \ldots + a_r \,.\, r \,.\, \rho_1^{-r} + \ldots = 0. \end{aligned}$$

Equating the coefficient of ρ_1^{-r} to zero gives—

$$ra_r + a_{r-1} = 0, \qquad \text{or} \qquad a_r = (-1)^r/r!$$

Thus a solution is

$$y(x) = a_0\left[1 - \frac{1}{1!}\frac{1}{x+1} + \frac{1}{2!}\frac{1}{(x+1)(x+2)} - \dots \right.$$
$$\left. - \frac{(-1)^r}{r!}\cdot\frac{1}{(x+1)(x+2)\dots(x+r)} + \dots\right],$$

a rapidly convergent series for x other than a negative integer.

EXERCISES (XV)

1. Solve the equation

$$ay(x+2) = (x+1)(x+2)\Delta y(x)$$

by expanding $y(x)$ as a series in ascending powers of a, and also by solving in inverse factorials.

2. Bessel's function of order n satisfies the difference equation

$$J_{n+1}(x) - 2nJ_n(x)/x + J_{n-1}(x) = 0.$$

Determine an expansion for $J_n(x)$ in ascending powers of x, given that the coefficient of the first term is $1/(2^n . n!)$.

3. Show that the equation

$$(a-1)(x+1)y(x+2) + (2x+1)y(x+1) - xy(x) = 0$$

can be written in the form—

$$(\pi^2 - a\rho^2)y(x) = 0$$

where $\pi = x\Delta$ and $\rho = xE$. Hence determine a solution in a factorial series.

4. Examine the difference equation

$$n^2 y_{n+2} + 3(n+2)y_{n+1} - 4y_n = 0$$

for a solution of the form—

$$y_n = \sum_{r=0}^{\infty} a_r/(n+r)!$$

and determine its convergence.

5. Express $x(x+1)y(x+2) - 2x(x+1)y(x+1) + (x^2-4)y(x) = 0$ in the form $F(\rho, \pi)y(x) = 0$. Hence obtain a solution in ascending factorials.

6. Find a solution of the equation—

$$(x+2)(x+1)y(x) - 2(x+2)^2 y(x+1) + (x+1)(x+3)y(x+2) = 0$$

in the form—

$$y(x) = a_0 + a_1 x + a_2 x(x-1)$$
$$+ \dots + a_r x(x-1)\dots(x-r+1) + \dots.$$

Examine the convergence of the series.

7. Obtain a solution of the equation

$$\lambda y(x) + (x - 1)^2 y(x - 1) - (x - 1)(x - 2)y(x - 2) = 0$$

in descending factorials in x.

Some General Properties of Certain Linear Difference Equations of the Second Order

We commence with some theorems relating to the behaviour of the solution of a difference equation

$$\Delta^2 u(x) + f(x + 1)u(x + 1) = 0 \tag{1}$$

as the function $f(x + 1)$ changes.

Comparison Theorem 1

Consider the two difference equations

$$\Delta^2 u(x) + f(x + 1)u(x + 1) = 0, \tag{1}$$

$$\Delta^2 v(x) + g(x + 1)v(x + 1) = 0, \tag{2}$$

where $g(x + 1) > f(x + 1) > 0$ for the range to be considered. We assume that over the range $0 \leqslant x < 2$, $u(x) = v(x)$, both being positive. We restrict ourselves to that range of x for which $u(x)$ and $v(x)$ continue to be positive.

We propose to show that $u(x) > v(x)$ as long as $u(x)$ and $v(x)$ remain positive.

Proof

Equations (1) and (2) may be rewritten—

$$u(x + 2) - 2u(x + 1) + u(x) = -f(x + 1)u(x + 1), \tag{3}$$

$$v(x + 2) - 2v(x + 1) + v(x) = -g(x + 1)v(x + 1). \tag{4}$$

Multiplying (3) by $v(x + 1)$, and (4) by $u(x + 1)$, and subtracting, we have

$$\Delta[u(x + 1)v(x) - u(x)v(x + 1)]$$
$$= [g(x + 1) - f(x + 1)]u(x + 1)v(x + 1). \tag{5}$$

For the range $0 \leqslant x < 1$, the right-hand side is positive so that,

$$\Delta[u(x + 1)v(x) - u(x)v(x + 1)] > 0,$$

and since $u(x+1)v(x+1)$ is positive for x in this range,

$$\frac{u(x+2)}{u(x+1)} - \frac{v(x+2)}{v(x+1)} > \frac{v(x)}{v(x+1)} - \frac{u(x)}{u(x+1)}.$$

The right-hand side is zero in this range since

$$u(x) = v(x), \qquad \text{and} \qquad u(x+1) = v(x+1).$$

Therefore, $u(x+2) > v(x+2)$. Thus for the range $0 \leqslant x < 2$, $u(x) = v(x)$ and for $2 \leqslant x < 3$, $u(x) > v(x)$. Now consider the range $1 \leqslant x < 2$. Here

$$u(x+1) > v(x+1), \qquad \text{and} \qquad u(x) = v(x),$$

therefore from (5)

$$\frac{u(x+2)}{u(x+1)} - \frac{v(x+2)}{v(x+1)} > 0.$$

Thus $u(x+2) > v(x+2)$, or for the range $3 \leqslant x < 4$, we have $u(x) > v(x)$.

Clearly this procedure may now be carried on so long as the functions $u(x)$ and $v(x)$ remain positive.

Example

We suppose x is a positive integer, and $f(x+1)$ and $g(x+1)$ are given by the following table

x	0	1	2	3	4	5	6
$f(x+1)$	0·05	0·10	0·15	0·20	0·25	0·30	0·35
$g(x+1)$	0·10	0·20	0·30	0·40	0·50	0·60	0·70

Suppose $\qquad u(0) = v(0) = 5, \qquad u(1) = v(1) = 6.$

The two equations may be written in the form

$$u(x+2) = [2 - f(x+1)]u(x+1) - u(x),$$
$$v(x+2) = [2 - g(x+1)]v(x+1) - v(x).$$

Inserting on the right the values of $u(x)$ and $v(x)$ at $x = 0$ and $x = 1$, and the corresponding values of $f(x+1)$ and $g(x+1)$, the values of $u(x)$ and $v(x)$ at $x = 2, 3, 4, 5, 6$ are easily calculated as follows—

x	0	1	2	3	4	5	6
$u(x)$	5	6	6·70	6·73	5·75	3·62	0·585
$v(x)$	5	6	6·40	5·52	2·98	− 1·36	

Here $v(x) \leqslant u(x)$ at least as far as the position at which $v(x)$ changes sign.

Corollary 1

If $M(x + 1) > f(x + 1) > m(x + 1)$, and if

$$\Delta^2u(x) + M(x + 1)u(x + 1) = 0,$$
$$\Delta^2v(x) + f(x + 1)v(x + 1) = 0,$$
$$\Delta^2w(x) + m(x + 1)w(x + 1) = 0,$$

then, provided that $u(x)$, $v(x)$, and $w(x)$ are identically equal over the range $0 \leqslant x < 2$, then $w(x) < v(x) < u(x)$, thereafter, as long as $u(x)$, $v(x)$, and $w(x)$ remain positive.

We note that if in the equation

$$\Delta^2v(x) + f(x + 1)v(x + 1) = 0$$

$f(x)$ is increased (or decreased) by a small positive quantity then $v(x)$, the solution, is decreased (or increased) by a positive amount. Again, if $f(x)$ is multiplied by a constant λ, then if $\lambda > 1$, the corresponding solution is less and when $\lambda < 1$ it is greater.

Corollary 2

Corollary 1 is valid up to the first zero of $w(x)$. When $u(x)$, $v(x)$, and $w(x)$ are continuous functions of x then the first zero of $v(x)$ falls between the first zeros of $u(x)$ and $w(x)$.

Corollary 3

The above theorem is valid whether $f(x)$ be positive or negative. When $f(x)$ is positive, as we have seen, this implies the delimitation of the position of the zeros. When $f(x)$ is negative, on the other hand, it can easily be shown, using the same theorem, that any solution of the equation lies between two functions which, if both have the same positive values between $x = 0$ and $x = 2$, provide upper and lower bounds to the actual solution, and which never become zero.

Between them, the two corollaries lead to the conclusion that when $f(x)$ is positive and lies between finite upper and lower bounds, the solution of the equation is an oscillating function with an unlimited number of zeros. When $f(x)$ is negative the solution increases with x, without limit.

If $u(x)$ and $v(x)$ are two independent solutions of a linear difference equation of the second order, then any other solution of the equation can be written in the form

$$y(x) = C_1(x)u(x) + C_2(x)v(x),$$

where $C_1(x)$ and $C_2(x)$ are unit periodics. All solutions obtained from any one particular solution by multiplying it by a unit periodic have the same zeros apart from those that may be introduced directly by the multiplying unit periodic. We define a fundamental solution as one which satisfies the equation and does not itself contain a unit periodic as a factor.

EXERCISE (XVI)

Show that if $y_1(x)$ is a solution of a second order difference equation containing a unit periodic, and $y(x)$ is a fundamental solution which is related to it, then

$$\frac{y(x+1)}{y(x)} = \frac{y_1(x+1)}{y_1(x)},$$

and $\dfrac{y(x+1)}{y(x)}$ does not contain a unit periodic. The solution of the first order difference equation

$$\frac{y(x+1)}{y(x)} = \frac{y_1(x+1)}{y_1(x)}$$

is a fundamental solution, where the arbitrary periodic that arises is written as unity.

Comparison Theorem 2

Let $u(x)$ and $v(x)$ be two fundamental solutions of the difference equation

$$\Delta^2 y(x) + \rho(x+1)y(x+1) = 0,$$

where $u(0) = 0$, $u(\eta) = 0$. We propose to show that $v(x)$ becomes zero at a point in the range $0 < x < \eta$.

Proof

$$\Delta^2 u(x) + \rho(x+1)u(x+1) = 0$$
$$\Delta^2 v(x) + \rho(x+1)v(x+1) = 0.$$

Eliminating $\rho(x + 1)$ leads to—

$$v(x + 1)\Delta^2 u(x) - u(x + 1)\Delta^2 v(x) = 0$$

or

$$\Delta[u(x)\Delta v(x) - v(x)\Delta u(x)] = 0.$$

Accordingly

$$u(x)\Delta v(x) - v(x)\Delta u(x) = c(x)$$

or

$$u(x)v(x + 1) - v(x)u(x + 1) = c(x),$$

where $c(x)$ is a unit periodic.

We assume that $u(x)$ and $v(x)$ are continuous functions, that $u(x)$ is positive, monotonic, and increasing for $0 < x < 2$, and that $v(x)$ is also positive, monotonic, but decreasing in that range. Hence $v(x + 1) < v(x)$ and $u(x + 1) > u(x)$ over the range $0 < x < 1$. Accordingly $c(x) < 0$ over the same range, and since $c(x)$ is a unit periodic it is therefore *everywhere* negative.

Now consider the situation at $x = \eta$ the first zero of $u(x)$. Here

$$u(\eta)v(\eta + 1) - v(\eta)u(\eta + 1) = c(\eta).$$

Since

$$u(\eta) = 0,$$

$$v(\eta) = -c(\eta)/u(\eta + 1).$$

Assume that $u(x)$ changes sign at its zero $x = \eta$, and that the next zero does not occur in the range $\eta < x < \eta + 1$, then $u(\eta + 1)$ is negative. Since $c(\eta)$ is also negative, it follows that $v(\eta)$ is negative. Hence $v(x)$ must have a zero in the range $2 < x < \eta$.

Characteristic Numbers and Characteristic or Eigen-functions of a Difference Equation Containing an Unspecified Constant λ

In this section we restrict ourselves to real integral values n of the variable x, and consider the equation

$$\Delta^2 u_n + \lambda f_{n+1} u_{n+1} = 0 \tag{1}$$

where f_{n+1} is a given function of n, finite in the range $0 < n < N$, N being a given positive integer.

Let $u_0 = 0$ and $u_N = 0$ be the boundary condition to be satisfied by the solution of the foregoing equation. This solution is of the form

$$u_n = Au_n(\lambda) + Bv_n(\lambda)$$

where A and B are arbitrary constants, and $u_n(\lambda)$ and $v_n(\lambda)$ are two independent solutions. Hence

$$0 = u_0 = Au_0(\lambda) + Bv_0(\lambda),$$
$$0 = u_N = Au_N(\lambda) + Bv_N(\lambda).$$

It follows that A and B are both zero so that u_n is zero everywhere, unless

$$0 = \begin{vmatrix} u_0(\lambda) & v_0(\lambda) \\ u_N(\lambda) & v_N(\lambda) \end{vmatrix}, \tag{2}$$

and this indicates that there may exist real values of λ, the eigen or characteristic values, to satisfy this relation, such that the corresponding equation

$$\Delta^2 u_n + \lambda f_{n+1} u_{n+1} = 0$$

has solutions, the eigen-functions, which vanish at $n = 0$ and $n = N$, but do not vanish everywhere. In fact for each such value, since A and B are not uniquely determined there exists a sheaf of solutions each member of which satisfies the necessary conditions of vanishing at $n = 0$ and $n = N$.

Equation (2), which provides the *characteristic values* λ, is however, expressed through the two basic solutions. This is a disadvantage since it makes it impossible to determine the values of λ without first obtaining the general solution of the difference equation. The values of λ can, however, be derived in another way, as follows. The equation states that

$$u_{n+2} = (2 - \lambda f_{n+1})u_{n+1} - u_n.$$

Hence $\quad u_2 = (2 - \lambda f_1)u_1,$

since $\quad u_0 = 0.$

$$u_3 = (2 - \lambda f_2)u_2 - u_1 = [(2 - \lambda f_1)(2 - \lambda f_2) - 1]u_1,$$
$$u_4 = (2 - \lambda f_3)u_3 - u_2 = \text{etc.},$$

and generally $u_N = F(\lambda)u_1$,
where $F(\lambda)$ is a polynomial of the $(N - 1)$th degree in λ. Hence the condition $u_N = 0$ requires $F(\lambda) = 0$ so that in general there are $(N - 1)$ values of λ corresponding to each of which there exists a difference equation (1) which possesses a sheaf of solutions vanishing at $n = 0$ and $n = N$.

Examples

$$\Delta^2 u_n + 2\lambda u_{n+1} = 0.$$

Let $u_0 = 0$ and $u_4 = 0$.

Here $u_{n+2} = 2(1 - \lambda)u_{n+1} - u_n$.

Hence

$$u_2 = 2(1 - \lambda)u_1,$$
$$u_3 = 2(1 - \lambda)u_2 - u_1,$$
$$u_4 = 2(1 - \lambda)u_3 - u_2.$$

Remembering that $u_4 = 0$, and eliminating u_1, u_2, u_3,

$$0 = \begin{vmatrix} 2(1-\lambda) & -1 & 0 \\ -1 & 2(1-\lambda) & -1 \\ 0 & -1 & 2(1-\lambda) \end{vmatrix},$$

giving $\lambda = 1,\ 1 - 1/\sqrt{2},\ 1 + 1/\sqrt{2}$ the three *characteristic values*. With these values of λ, we solve the original equation in each case, and determine the eigen-functions.

EXERCISE (XVII)

By solving the equation

$$\Delta^2 u_n + 2\lambda u_{n+1} = 0,$$

subject to the conditions $u_0 = u_N = 0$, show that the characteristic values of λ are $2 \sin^2 \theta/2$, where

$$\theta = \pi/N,\ 2\pi/N,\ \ldots,\ (N-1)\pi/N.$$

Theorem

If U_n and V_n be two characteristic functions corresponding to the two characteristic numbers λ_r and λ_s then U_n and V_n satisfy the *orthogonal conditions*—

$$\sum_{n=0}^{n=N-1} f_{n+1} U_{n+1} V_{n+1} = 0.$$

To establish this we note that

$$\Delta^2 U_n + \lambda_r f_{n+1} U_{n+1} = 0,$$
$$\Delta^2 V_n + \lambda_s f_{n+1} V_{n+1} = 0,$$

where $\lambda_r \neq \lambda_s$, $U_0 = V_0 = 0$, $U_N = V_N = 0$, and $U_n \neq A V_n$ where A is a constant. Multiplying the first of the two difference equations by V_{n+1} and the second by U_{n+1} and subtracting we have

$$-(\lambda_r-\lambda_s)f_{n+1}U_{n+1}V_{n+1}=V_{n+1}\Delta^2U_n-U_{n+1}\Delta^2V_n$$
$$=\Delta[U_{n+1}V_n-U_nV_{n+1}].$$

Summing both sides from $n=0$ to $n=N-1$

$$-(\lambda_r-\lambda_s)\sum_0^{N-1}f_{n+1}U_{n+1}V_{n+1}=[U_{n+1}V_n-U_nV_{n+1}]_0^{N-1}$$
$$=(U_NV_{N-1}-U_{N-1}V_N)-(U_1V_0-U_0V_1)$$
$$=0,$$

since $U_N=0=V_N=U_0=V_0$.

Since $\lambda_r\neq\lambda_s$ it follows that

$$\sum_0^{N-1}f_{n+1}U_{n+1}V_{n+1}=0.$$

Expansion in Eigen-functions

A function W_n assumes the $N-1$ arbitrary finite values W_1, $W_2, \ldots, W_{N-1}$ at $n=1, 2, \ldots, N-1$. It is clearly possible to express W_n as a linear combination of any given $N-1$ functions of n say p_n, q_n, r_n etc., i.e. to find $A_1, A_2, \ldots, A_{N-1}$ such that the values of $A_1p_n+A_2q_n+A_3r_n+\ldots+A_{N-1}z_n$ at $n=1, 2, \ldots, N-1$ are equal respectively to $W_1, W_2, \ldots, W_{N-1}$. This merely requires the solution of a set of $N-1$ linear equations. If, however, the functions p_n, q_n, r_n etc. are the characteristic or eigen-functions of a difference equation of the type under consideration, the coefficients $A_1, A_2, \ldots, A_{N-1}$ can be expressed in simple form.

Let $U_n(\lambda_r)$ be the eigen-function corresponding to the eigen-value λ_r of which as we have seen there are $N-1$ in all, and write

$$W_n=\sum_1^{N-1}A_rU_n(\lambda_r).$$

Hence multiplying both sides by $f_{n+1}U_n(\lambda_r)$ and summing from $n=1$ to $n=(N-1)$ we have

$$\sum_{n-1}^{n=N-1}f_{n+1}U_n(\lambda_r)W_n=\sum_{n=1}^{N-1}[A_1U_n(\lambda_1)+A_2U_n(\lambda_2)+\ldots$$
$$+A_{N-1}U_n(\lambda_{N-1})]f_{n+1}U_n(\lambda_r).$$

Now
$$\sum_{n=1}^{N-1} U_n(\lambda_r)U_n(\lambda_1)f_{n+1} = 0,$$
$$\sum_{n=1}^{N-1} U_n(\lambda_2)U_n(\lambda_r)f_{n+1} = 0,$$

and so on.

The only summation which does not vanish is

$$\sum_{1}^{N-1} U_n(\lambda_r)U_n(\lambda_r)f_{n+1} = \sum_{1}^{N-1} U_n(\lambda_r)^2 f_{n+1}.$$

Hence
$$\sum_{n=1}^{N-1} f_{n+1}U_n(\lambda_r)W_n = A_r \sum_{1}^{N-1} U_n(\lambda_r)^2 f_{n+1}$$

and this consequently determines the coefficient A_r explicitly.

Because W_n is here expressed in terms of eigen-functions which themselves separately vanish at $n = 0$ and $n = N$, the expression for W_n also vanishes at these values of n.

Rayleigh's Principle Applied to Linear Difference Equations —Approximate Determination of the Eigen-values

If the equation

$$\Delta^2 U_n + \lambda f_{n+1} U_{n+1} = 0$$

be multiplied by U_{n+1}, then

$$U_{n+1}\Delta U_{n+1} - U_n \Delta U_n = -\lambda f_{n+1} U_n^2 + (\Delta U_n)^2.$$

Hence summing both sides from $n = 0$ to $n = N$ and making $U_0 = U_N = 0$ we obtain

$$\sum_{0}^{N-1} (\Delta U_n)^2 = \lambda \sum_{0}^{N-1} f_{n+1} U_{n+1}^2.$$

This is a relation which holds between the eigen-values λ and the eigen-functions U_n, from which it follows that provided U_n is real then λ is positive.

In general, of course, U_n is not directly obtainable. If, however, in the above summation we replace U_n by a known function p_n which is zero at $n = 0$ and $n = N$ it is possible to approximate quite closely to λ.

Consider the identity $\Delta[V_n \Delta U_n] = V_{n+1}\Delta^2 U_n + \Delta U_n \cdot \Delta V_n$, and let $V_n = p_n^2/U_n$. Then—

$$\begin{aligned}\Delta[p_n^2(\Delta U_n)/U_n]\\ &= p_{n+1}^2(\Delta^2 U_n)/U_{n+1} + (U_{n+1} - U_n)(p_{n+1}^2/U_{n+1} - p_n^2/U_n)\\ &= p_{n+1}^2(\Delta^2 U_n)/U_{n+1} + (p_{n+1} - p_n)^2 - U_n U_{n+1}(p_{n+1}/U_{n+1} - p_n/U_n)^2.\end{aligned}$$

But $\Delta^2 U_n = -\lambda f_{n+1} U_{n+1}$, and consequently

$$\Delta[p_n^2(\Delta U_n)/U_n] = -\lambda f_{n+1} p_{n+1}^2 + (\Delta p_n)^2 - U_n U_{n+1}[\Delta(p_n/U_n)]^2.$$

If both sides be now summed from $n = 0$ to $n = N - 1$, bearing in mind that $U_0 = U_N = p_0 = p_N = 0$ we get

$$\lambda \sum_0^{N-1} f_{n+1} p_{n+1}^2 = \sum_0^{N-1} (\Delta p_n)^2 - \sum_0^{N-1} U_n U_{n+1}[\Delta(p_n/U_n)]^2.$$

Since $n = N$ is presumed to be the first zero of pn and U_n, up to that position U_n and U_{n+1} have the same sign so that the last summation on the right is essentially positive.

Thus
$$\lambda < \sum_0^{N-1} (\Delta p_n)^2 / \sum_0^{N-1} f_{n+1} p_{n+1}^2.$$

This is the analogue to Rayleigh's theorem in relation to eigen-values of a second order differential equation. It shows that if an approximation p_n of any kind be taken to be the solution of the difference equation of such a nature that the boundary conditions are satisfied, and these summations may be conveniently evaluated, then an upper bound to the lowest eigen-value is immediately determined.

Example

$$\Delta^2 u_n + 2\lambda u_{n+1} = 0,$$

where
$$U_0 = U_4 = 0.$$

Take as an approximate solution p_n where

n	0		1		2		3		4
p_n	0		1		2		1		0
Δp_n		1		1		-1		-1	

Here
$$f_{n+1} = 2$$

$$\sum_0^{N-1} (\Delta p_n)^2 = 1 + 1 + 1 + 1 = 4,$$

$$\sum_{0}^{N-1} f_{n+1}p_{n+1}^2 = 2(1 + 4 + 1) = 12.$$

Hence
$$\lambda < 4/12,$$
i.e.
$$\lambda < 0{\cdot}333.$$

But we have already found that the lowest value of λ is $1 - \cos\theta$ where
$$\theta = \pi/N = \pi/4,$$
i.e.
$$\lambda = 1 - \cos\pi/4 = 1 - 1/\sqrt{2} = 0{\cdot}29$$

which is consistent with the above theorem.

This, however, gives only an upper bound to λ. Since in general $0 < m < f_{n+1} < M$, we may say that the solution of

$$\Delta^2 U_n + \lambda f_{n+1} U_{n+1} = 0,$$

lies between those of

$$\Delta^2 V_n + \lambda M V_{n+1} = 0,$$
and
$$\Delta^2 W_n + \lambda m W_{n+1} = 0,$$

provided $V_0 = W_0 = 0$ and $V_1 = W_1$. Thus it is to be expected that N, the position at which U_n is zero, lies between the first zero of V_n and that of W_n. Since the corresponding equations have constant coefficients, these zeros are easily determined.

Now the solution of

$$\Delta^2 V_n + \lambda M V_{n+1} = 0,$$

subject to $V_0 = 0$ and $V_1 = V_1$ is,

$$V_n = \frac{2V_1}{\sqrt{(4\lambda M - \lambda^2 M^2)}} \sin n\theta$$

where $2\sin\theta = \sqrt{(4\lambda M - \lambda^2 M^2)}$ and this would change sign next where

$$N \fallingdotseq \frac{\pi}{\sin^{-1}\frac{1}{2}\sqrt{(4\lambda M - \lambda^2 M^2)}}.$$

Similarly
$$N \fallingdotseq \frac{\pi}{\sin^{-1}\frac{1}{2}\sqrt{(4\lambda m - \lambda^2 m^2)}}.$$

Accordingly we may write

$$\frac{\pi}{\sin^{-1} \frac{1}{2}\sqrt{(4\lambda M - \lambda^2 M^2)}} < N < \frac{\pi}{\sin^{-1} \frac{1}{2}\sqrt{(4\lambda m - \lambda^2 m^2)}},$$

or

$$\frac{4}{M} \sin^2 \frac{\pi}{2N} < \lambda < \frac{4}{m} \sin^2 \frac{\pi}{2N}$$

giving upper and lower bounds to λ. If Rayleigh's theorem gives a lower value for the upper bound it is preferable to the upper bound as given above.

EXERCISES (XVIII)

1. In each of the following equations $y(0) = 0$, Determine approximately the position where $y(n)$ changes sign—

(i) $\Delta^2 y_n + (n^2 + 11)y_{n+1}/(n^2 + 10) = 0.$
(ii) $\Delta^2 y_n + (1 + e^{-n})y_{n+1} = 0.$
(iii) $(n^2 + 20)\Delta^2 y_n + (2n^2 + 13)y_{n+1} = 0.$

2. Determine upper and lower bounds to the lowest characteristic value λ for the equation

$$\Delta^2 y_n + \lambda[n^2 + n + 5)/(n^2 - n + 5)]y_{n+1} = 0$$

given $y_0 = 0$ and $y_5 = 0$.

CHAPTER 7

Some Applications of Difference Equations

THIS chapter contains a miscellaneous collection of problems in the solution of which the theory of difference equations is exemplified.

1. *A Problem of Continuous Beams.* A light beam of length Na units is simply supported at N points an equal distance a, apart as shown in the diagram. A weight P is placed at the origin. We require the bending moments at the supports.

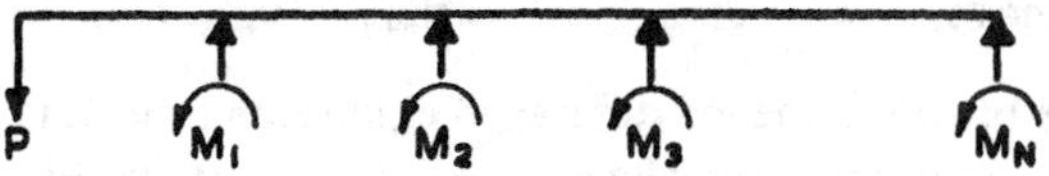

FIG. 7.1

The moments at the successive supports satisfy the Equation of Three Moments, viz.

$$M_{n+2} + 4M_{n+1} + M_n = 0, \tag{1}$$

with the conditions $M_1 = Pa$, $M_N = 0$. Equation (1) is, in operator notation,

$$(\mathrm{E}^2 + 4\mathrm{E} + 1)M_n = a,$$

and its general solution may be written

$$M_n = A(-2 + \sqrt{3})^n + B(-2 - \sqrt{3})^n,$$

where A and B are arbitrary constants.

Thus $\qquad -Pa = A(-2 + \sqrt{3}) + B(-2 - \sqrt{3})$

$$0 = A(-2 + \sqrt{3})^N + B(-2 - \sqrt{3})^N.$$

So that $\qquad M_n = -Pa[\lambda^{n-N} - \mu^{n-N}]/[\lambda^{1-N} - \mu^{1-N}]$

where $\qquad \mu = -2 + \sqrt{3}$

and $\qquad \lambda = -2 - \sqrt{3}.$

When the beam is of infinite extent so that $N \to \infty$,

$$M_n \to Pa(-1)^n[2 - \sqrt{3}]^{n-1}.$$

2. Two spheres of masses M and m, $(M > m)$, and coefficient of restitution e, lie on a smooth horizontal surface with the line of centres at right-angles to two walls which are perfectly elastic. M is projected at m with velocity U while m is initially at rest. We require the velocities just before the nth impact between the balls.

Let u_n and v_n be the velocities of M and m just before the nth impact. The velocities of M and m just before the $(n+1)$th impact are u_{n+1} and v_{n+1}; thus the velocities just after the nth impact are u_{n+1} and v_{n+1} since the wall is perfectly elastic.

The momentum equation is

$$Mu_n + mv_n = -mv_{n+1} + Mu_{n+1}.$$

Also we have, $$-e(u_n - v_n) = u_{n+1} + v_{n+1}.$$

This pair of simultaneous difference equations in u_n and v_n together with the boundary conditions, $u_1 = U$, $v_1 = 0$, enable us to solve for u_n and v_n completely. Eliminating v_n we obtain,

$$\left[\mathrm{E}^2 + \frac{(m - M)(e + 1)}{M + m}\mathrm{E} + e\right]u_n = 0,$$

whence u_n can be obtained as a function of n; v_n can then be obtained directly from the equation

$$m(e + 1)v_n = (M + m)u_{n+1} - (M - me)u_n.$$

For instance, if $M = 4m$ and $r = 1$, the equations to determine u_n and v_n become,

$$[\mathrm{E}^2 - 6\mathrm{E}/5 + 1]u_n = 0,$$

and $$2v_n = 5u_{n+1} - 3u_n.$$

Hence, $$u_n = A[(3 + 4i)/5]^n + B[(3 - 4i)/5]^n$$
$$= C \sin (n\theta + \alpha),$$

where $$\tan \theta = 4/3$$

and C and α are arbitrary constants. Hence

$$2v_n = 5C \sin (n\theta + \theta + \alpha) - 3C \sin (n\theta + \alpha).$$

The boundary conditions require

$$U = C \sin(\theta + \alpha)$$
$$0 = 5 \sin(2\theta + \alpha) - 3 \sin(\theta + \alpha),$$

from which we obtain,

$$U = C$$

and $$\alpha = \tfrac{1}{2}\pi - \theta = \tfrac{1}{2}\pi - \tan^{-1} 4/3.$$

Thus $$u_n = U \cos[(n-1)\theta]$$

and $$v_n = \frac{3U}{2} \cos[(n-1)\theta] - \frac{5U}{2} \cos n\theta.$$

We notice that although u_n and v_n are expressed in terms of trigonometrical functions, u_n and v_n are not periodic since $\tan^{-1} 4/3$ is not an integral fraction of 2π.

EXERCISE (I)

If $e = 1$, determine for what mass ratios, if any, the velocity configuration will repeat.

3. Two balls of coefficient of restitution e and masses M and m are set running in a circular groove in a plane surface with velocities U and V where $U > V$. What are the velocities of the balls just before the nth impact?

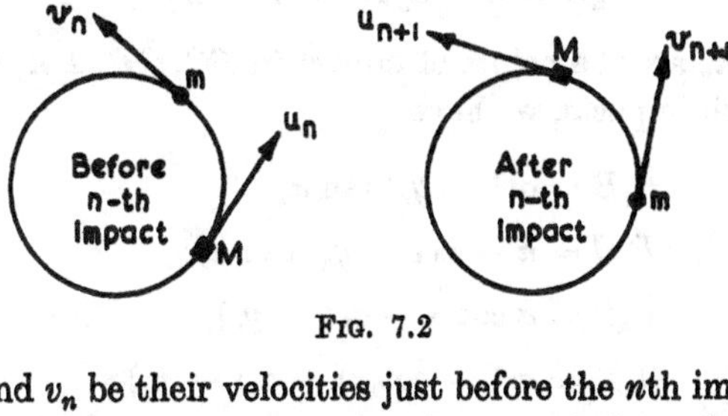

FIG. 7.2

Let u_n and v_n be their velocities just before the nth impact. Then since u_{n+1} and v_{n+1} are the velocities just before the $(n+1)$th impact these must also be their velocities just after the nth impact.

$$\text{Momentum before} = Mu_n + mv_n,$$
$$\text{momentum after} = Mu_{n+1} + mv_{n+1}.$$

Hence, $$Mu_n + mv_n = Mu_{n+1} + mv_{n+1}.$$

Accordingly $$A = Mu_n + mv_n = MU + mV,$$

where A is a constant fixed in terms of the initial velocities. Again, we have,

relative velocity before the nth impact $= u_n - v_n$,
relative velocity after the nth impact $= v_{n+1} - u_{n+1}$.

Thus $$-e(v_n - u_n) = v_{n+1} - u_{n+1}.$$

Hence $$v_n - u_n = B(-1)^n e^n$$

and $$V - U = -Be$$

or, $$v_n - u_n = (U - V)(-1)^n e^{n-1}.$$

Hence, $$u_n = [MU - mV - m(U - V)(-1)^n e^{n-1}]/(M + m)$$
$$v_n = [MU - mV + M(U - V)(-1)^n e^{n-1}]/(M + m)$$

Now since $e < 1$, $e^n \to 0$ as $n \to \infty$. Hence,

$$\lim_{n\to\infty} u_n = \lim_{n\to\infty} v_n = [MU + mV]/(M + m).$$

4. *ABCD* is a square billiard table of side *a*. A billiard ball is projected from a point *O* in *AB* making an angle α with *AB* and is reflected on *BC*, *CD*, *DA* in succession. Where does it strike *AB* after the *n*th circuit?

Let O_n and O_{n+1} be the point of impact in AB after the nth and the $(n + 1)$th circuit. Let $ay_n = AO_n$, then $ay_{n+1} = AO_{n+1}$. Then we have,

$$O_nB = a - ay_n = a(1 - y_n)$$

If P_n, Q_n, R_n are the points of impact in BC, CD, DA, respectively after the nth impacts, we have,

$$P_nB = a(1 - y_n)\tan\alpha,$$
$$P_nC = a - a(1 - y_n)\tan\alpha,$$
$$Q_nC = a\cot\alpha - a(1 - y_n),$$
$$Q_nD = a(1 - \cot\alpha) + a(1 - y_n),$$
$$R_nD = a(\tan\alpha - 1) + a(1 - y_n)\tan\alpha,$$
$$R_nA = a(2 - \tan\alpha) - a(1 - y_n)\tan\alpha,$$
$$AO_{n+1} = a(2\cot\alpha - 1) - a(1 - y_n) = ay_{n+1}.$$

Thus $$y_{n+1} = y_n + 2(\cot\alpha - 1),$$

and $$y_n = y_0 + 2n(\cot\alpha - 1)$$

where $AO = ay_0$ the starting point, a solution that will be valid until the ball no longer strikes all four sides in the order indicated. As long as it continues to do so, however, the point of impact moves along in one direction always by equal amounts, except when $\alpha = \pi/4$ when it returns on itself.

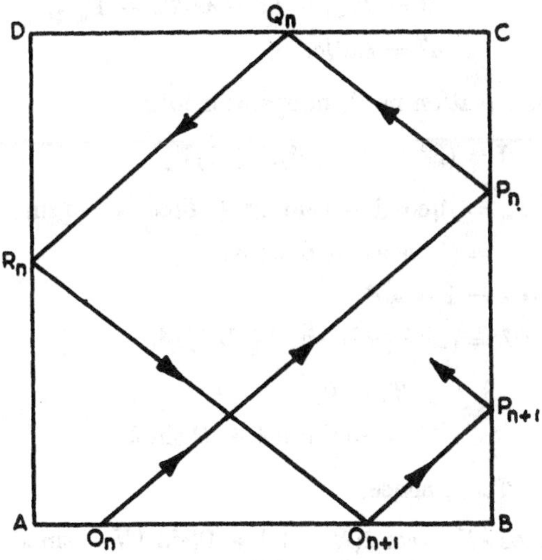

FIG. 7.3

5. Consider a uniform shaft of constant cross section and carrying N identical discs spaced at equal intervals l.

Let the torsional stiffness of the shaft be c, and θ_n the angular displacement of the nth disc relative to the first. Thus the torque to the left of this disc is $c(\theta_{n-1} - \theta_n)$ and that to the right is $c(\theta_n - \theta_{n+1})$ acting, however, on the nth disc in opposite directions. Hence, if I is the moment of inertia of the disc,

$$I\frac{d^2\theta_n}{dt^2} = c(\theta_{n-1} - \theta_n) - c(\theta_n - \theta_{n+1})$$

$$= c(\theta_{n+1} - 2\theta_n + \theta_{n-1})$$

$$= c\Delta^2\theta_{n-1}.$$

We proceed to examine what torsional vibrations may be set up when one end of the shaft is held rigid at the first disc, i.e. $\theta_1 = 0$. At the other end, which is free, the torque is zero so that $\theta_n = \theta_{n-1}$.

Let $\theta_n = T_n \sin \omega t$ represent the motion of the discs, then, inserting into the difference-differential equation,

$$-I\omega^2 T_n = c\Delta^2 T_{n-1}$$

where $\qquad T_1 = 0 \quad \text{and} \quad T_N = T_{N-1}.$

Thus $\qquad 0 = T_{n+1} - (2 - \alpha^2)T_n + T_{n-1},$

where $\qquad \alpha^2 = I\omega^2/c.$

Writing the equation in T_n in operator form,

$$0 = (\mathrm{E}^2 - (2 - \alpha^2)\mathrm{E} + 1)T_n.$$

Hence $\qquad T_n = A[\cos \lambda + \mathrm{i} \sin \lambda]^n + B[\cos \lambda - \mathrm{i} \sin \lambda]^n$
$\qquad\qquad = C \cos n\lambda + D \sin n\lambda,$

where $\quad \cos \lambda = 1 - \frac{1}{2}\alpha^2$

or $\qquad \lambda = \sin^{-1} \frac{1}{2}\alpha = \sin^{-1} \frac{1}{2}(I\omega^2/c)^{\frac{1}{2}}.$

Now, $\qquad T_1 = 0,$

hence, $\qquad 0 = C \cos \lambda + D \sin \lambda.$

Also, $T_N = T_{N-1}$, hence,

$$0 = C[\cos \lambda N - \cos \lambda(N-1)] + D[\sin \lambda N - \sin \lambda(N-1)];$$

thus C and D are zero unless

$$0 = \cos \lambda[\sin \lambda N - \sin \lambda(N-1)] - \sin \lambda[\cos \lambda N - \cos \lambda(N-1)]$$

giving $\qquad \lambda = 2\pi,$

i.e. $\qquad \frac{1}{2}\alpha = \sin \lambda = 0,$

which gives no vibrations,

or $\qquad 0 = (N-1)\lambda - \pi + \lambda(N-2),$

i.e. $\qquad \lambda = \pi/(2N-3),$

for the slowest oscillation, so that

$$\omega(I/c)^{\frac{1}{2}} = 2 \sin [\pi/(2N-3)].$$

6. *Voltage Drop in a Chain of Insulators.* For continuity in the currents (*see* Fig. 7.4) we require,

$$I_n + i_n = I_{n+1}, \tag{1}$$

$$C_1 \frac{d}{dt}(V_n - V_{n-1}) = -I_n. \tag{2}$$

Also

$$c_2 \frac{dV_n}{dt} = -i_n, \tag{3}$$

where V_n is the voltage of a line conductor.

We suppose that V_n is an alternating voltage of frequency ω, say $W_n \cos 2\pi\omega t$. Hence,

$$V_n = W_n \cos 2\pi\omega t.$$

and so from (2) and (3),

$$I_n = J_n \sin 2\pi\omega t \tag{4}$$

and

$$i_n = j_n \sin 2\pi\omega t. \tag{5}$$

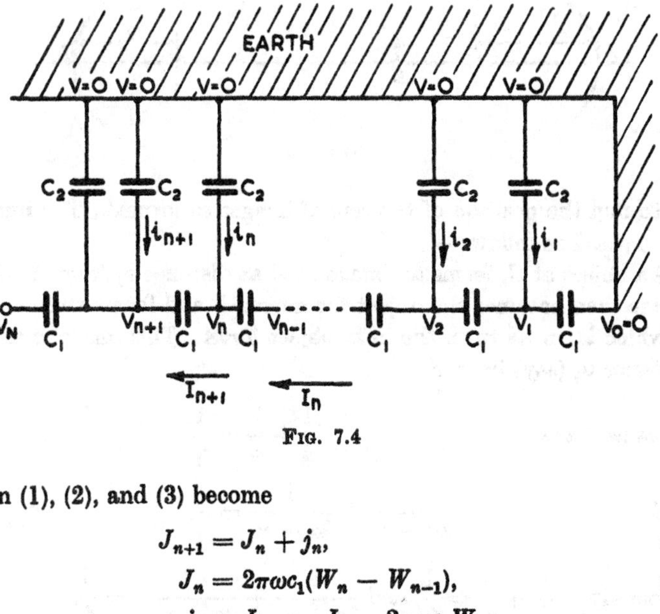

FIG. 7.4

Then (1), (2), and (3) become

$$J_{n+1} = J_n + j_n, \tag{6}$$

$$J_n = 2\pi\omega c_1(W_n - W_{n-1}), \tag{7}$$

$$j_n = J_{n+1} - J_n = 2\pi\omega c_2 W_n. \tag{8}$$

Equations (7) and (8) can be written, on raising n to $(n + 1)$

$$2\pi\omega c_1(\mathrm{E} - 1)W_n - \mathrm{E}J_n = 0,$$

$$2\pi\omega c_2 W_n - (\mathrm{E} - 1)J_n = 0.$$

Hence, eliminating J_n operationally,

$$[c_1(\mathrm{E} - 1)^2 - c_2\mathrm{E}]W_n = 0$$

or

$$[\mathrm{E}^2 - (2 + c_2/c_1)\mathrm{E} + 1]W_n = 0.$$

The conditions to be satisfied are $W_0 = 0$ and W_N given, from which it easily follows that

$$W_n = W_N \sinh \lambda n / \sinh \lambda N,$$

where $\cosh \lambda = 1 + c_2/2c_1$ or $\sinh \frac{1}{2}\lambda = \frac{1}{2}\sqrt{(c_2/c_1)}$.

Thus the voltage at any position n is given by

$$V_n = W_N \cos 2\pi\omega t \sinh \lambda n / \sinh \lambda N.$$

7. An object is placed between two concave mirrors A and B of focal lengths f_1 and f_2 respectively. The mirrors are a distance $2a$ apart. The object is placed at C, a distance u from A.

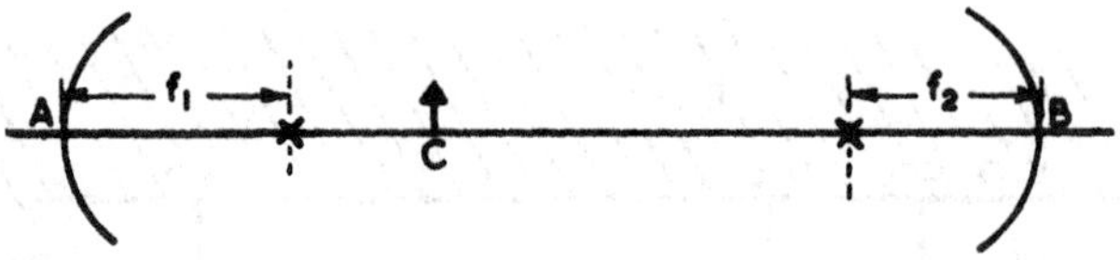

FIG. 7.5

To find the position of the sets of images so formed. The images are created as follows—

An object at C, forms an image in A at distance v_1 from A. This image then becomes the object for mirror B and forms an image in it which is, in its turn, the next object for A. This image is now a distance u_2 (say) from A.

Thus we have $$\frac{1}{u} + \frac{1}{v_1} = \frac{1}{f_1},$$

and $$\frac{1}{2a - v_1} + \frac{1}{2a - u_2} = \frac{1}{f_2}.$$

Generally, $\dfrac{1}{u_n} + \dfrac{1}{v_n} = \dfrac{1}{f_1}$, and $\dfrac{1}{2a - v_n} + \dfrac{1}{2a - u_{n+1}} = \dfrac{1}{f_2}$.

Eliminating v_n between these two equations we obtain,

$$u_{n+1} = \frac{u_n[4af_2 - f_1f_2 + 2af_1 - 4a^2] + 4a[af_1 - f_1f_2]}{u_n(f_1 + f_2 - 2a) - f_1f_2 + 2af_1}.$$

This is an equation of the form,

$$u_{n+1} = \frac{au_n + b}{cu_n + d},$$

which can be solved immediately by writing $u_n = r_n/s_n$ and a substitution made to find v_n. The solution contains one constant which can be determined since $u_n = u$ when $n = 1$.

In general, the solution of this type of equation is of the form,

$$u_n = \frac{r_n}{s_n} = \frac{A[\lambda/\mu]^n + B}{A_1[\lambda/\mu]^n + B_1}$$

where A_1 and B_1 are known in terms of A and B, and λ and μ are the roots of the equation

$$z^2 - (a + d)z + bc + ad = 0.$$

A special case arises where $\lambda = -\mu$, i.e. $a + d = 0$, i.e. in terms of the quantities given

$$f_1 f_2 = 2a[f_1 + f_2 - a].$$

In this case u_n is a periodic function of period two in n and there are only four distinct images formed in the two mirrors, two distinct positions for each mirror, and the alternate images fall on each other.

8. A series of masses $m_1, m_2, \ldots, m_n, \ldots, m_N$ are fixed at evenly spaced intervals l along a light string of length $(N + 1)l$, subjected to a tension T_0. The whole system is set rotating with angular velocity Ω about the axis of the string. It is required to find the values of Ω that correspond to an unchanging configuration, slightly displaced from the initial straight position.

Let T_n and T_{n+1} be the tensions in the sections of string to the left and to the right of the mass m_n, and θ_n and θ_{n+1} the slopes in these sections. Let y_n be the displacement of m_n from the axis, then

Radially $$m_n y_n \Omega^2 = T_n \sin\theta_n - T_{n+1}\sin\theta_{n+1}$$
$$= -\Delta(T_n\theta_n),$$

since θ_n is small.

Horizontally $$T_n \cos\theta_n = T_{n+1}\cos\theta_{n+1},$$

i.e. $$T_n = T_{n+1} - T_0.$$

Thus $$m_n y_n \Omega^2 = -T_0 \Delta\theta_n.$$

Now $$\theta_n = \Delta y_{n-1}/l,$$

so that $$\Delta^2 y_{n-1} + l\Omega^2 m_n y_n/T_0 = 0,$$

or $$\Delta^2 y_n + l\Omega^2 m_{n+1} y_{n+1}/T_0 = 0.$$

Let $m_n = m_1 M_n$ so that M_n is a pure number, then

$$\Delta^2 y_n + \lambda M_{n+1} y_{n+1} = 0,$$

where $$\lambda = l\Omega^2 m_1/T_0.$$

This is an equation for y_n, subject to the boundary conditions that $y_0 = 0$ and $y_{N+1} = 0$. This can give a non-zero value for y_n only for a set of eigen-values of λ (*see* Chapter 6) corresponding to each one of which there exists a special value of Ω and a special configuration of the string.

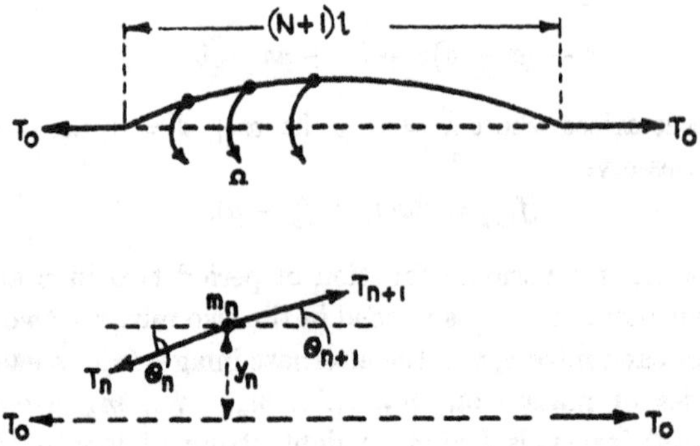

FIG. 7.6

Special Cases

(*a*) $M_n = 1$. All the masses are equal, and the difference equation is

$$\Delta^2 y_n + \lambda y_{n+1} = 0; \quad y_0 = 0, \ y_{N+1} = 0.$$

The general solution is—

$$y_n = A \cos n\theta + B \sin n\theta,$$

where $$\lambda = 4 \sin^2 \theta/2.$$

The conditions $y_0 = 0$ and $y_{N+1} = 0$ require $A = 0$, B arbitrary and $(N + 1)\theta = r\pi$, if y_n is not to be identically zero, where $r = 1, 2, 3 \ldots$ Accordingly

$$y_n = B \sin n\theta = B \sin [nr\pi/(N + 1)],$$

and $$\lambda = 4 \sin^2 \theta/2 = 4 \sin^2 [r\pi/2(N + 1)]$$

so that $$\Omega = (4T_0/m_1 l)^{\frac{1}{2}} \sin r\pi/2(N + 1).$$

The case $r = 1$ represents the configuration where all the masses are situated simultaneously on one side of the axis of rotation. With

$r = 2$ the masses lie on a sine curve rotating about the axis and crossing the latter mid-way between the two ends of the string.

(*b*) $M_1 = M_5 = 1,\ M_2 = M_4 = 4,\ M_3 = 6.$

Here $N = 5$ and we approximate to λ by using the extension of Rayleigh's principle developed in Chapter 6. Let the approximate shape of the string be $p_n = \sin(\pi n/6)$ which is zero at the terminal points $n = 0$ and $n = 6$. Then

$$\lambda \leqslant \sum_1^5 (\Delta p_n)^2 / \sum_1^5 M_n p_n^2$$

$$\Delta p_n = 2\cos(\pi n/6 + \pi/12)\sin \pi/12,$$

$$\sum_1^5 (\Delta p_n)^2 = 3(2 - \sqrt{3}),$$

$$\sum_1^5 M_n p_n^2 = \sin^2(\pi/6) + 4\sin^2(\pi/3) + 6\sin^2(\pi/2) + 4\sin^2(2\pi/3) + \sin^2(5\pi/6)$$

$$= 1/4 + 3 + 6 + 3 + 1/4 = 12{\cdot}5.$$

Hence the lowest value of Ω is approximately

$$\Omega \doteqdot \sqrt{[6(2 - \sqrt{3})T_0/25m_1 l]}.$$

9. Into how many separate compartments will n non-parallel lines in a plane divide that plane?

Let u_n be this number. The $(n + 1)$th line will be cut by the n lines in n points and will therefore divide $n + 1$ previously existing compartments into two. Thus

$$u_{n+1} = u_n + n + 1,$$

i.e.
$$u_n = (n + 1)n/2 + A.$$

When $n = 1$, $u_1 = 2$ so that $A = 1$. Hence finally—

$$u_n = (n^2 + n + 2)/2.$$

10. Into how many non-overlapping compartments will n non-concurrent planes divide three-dimensional space?

Let u_n be the number of such compartments. Take an $(n + 1)$th plane to cut the previous n planes. It will itself be cut in n lines which are non-concurrent, since this $(n + 1)$th plane cannot itself pass through the common point of any three of the previous planes.

These n lines divide this $(n+1)$th plane in $(n^2+n+2)/2$ non-overlapping regions. Each of these regions represents the plane surface of division across a three-dimensional compartment that arose from the previous n planes. Thus the $(n+1)$th plane adds $(n^2+n+2)/2 = 1 + \frac{1}{2}n(n+1)$ new compartments, or

$$u_{n+1} = u_n + 1 + n(n+1)/2,$$

Hence $u_n = A + n + (n-1)n(n+1)/6$.

When $n = 1$, $u_1 = 2$, so that $A = 1$.

Thus $u_n = (n+1)[1 + n(n-1)/6] = (n+1)(n^2 - n + 6)/6$.

Check

$u_2 = 4$, $u_3 = 8$, as with the co-ordinate planes.

11. In how many ways can n parcels and n labels be so muddled that no parcel has its own label?

Let U_n be the number of ways. We may suppose that the labels are compartments into which the parcels are supposed to be placed.

Now U_n will be composed of two groups—

1. One in which two parcels, say C and D merely interchange compartments.

2. Those in which C goes to D's compartment but D does not go to C's.

For (1), once C is fixed, D can be chosen in $(n-1)$ ways and when interchanged the rest are allowed to get muddled in U_{n-2} ways.

For (2), we place C in any other compartment in $(n-1)$ ways and allow the rest to be muddled but barring D from C's compartment. This means that in the muddling D behaves as if it were C. Hence the total number here is $(n-1)U_{n-1}$. Thus,

$$U_n = (n-1)U_{n-1} + (n-1)U_{n-2}$$

or

$$0 = U_{n+2} - (n+1)U_{n+1} - (n+1)U_n.$$

Solution

Let R_{n+2} be a factor that makes this exact. Then

$$\begin{aligned} 0 &= R_{n+2}U_{n+2} - (n+1)R_{n+2}U_{n+1} - (n+1)R_{n+2}U_n \\ &= (R_{n+2}U_{n+2} - R_{n+1}U_{n+1}) + [R_{n+1} - (n+1)R_{n+2}]U_{n-1} \\ &\qquad - (n+1)R_{n+2}U_n \\ &= \Delta(R_{n+1}U_{n+1}) + \Delta[(R_n - nR_{n+1})U_n] \\ &\qquad + [R_n - nR_{n+1} - (n+1)R_{n+2}]U_n. \end{aligned}$$

Hence we require,

$$(n+1)R_{n+2} + nR_{n+1} - R_n = 0.$$

A solution of this equation is $R_n = (-1)^n$. Hence the difference equation can be transformed to

$$0 = \Delta(R_{n+1}U_{n+1}) + \Delta[(R_n - nR_{n+1})U_n],$$

where $\quad R_n = (-1)^n,$

or $\quad A(-1)^n = U_{n+1} - (n+1)U_n,$

where A is an arbitrary constant.

Now obviously $U_1 = 0$

and $\quad U_2 = 1,$

therefore $\quad 1 = -A.$

This gives, $\quad 0 = U_{n+1} - (n+1)U_n - (-1)^{n+1}.$

To solve this first order equation in U_n, we divide through by $(n+1)!$ and obtain

$$\Delta(U_n/n!) = (-1)^{n+1}/(n+1)!$$

$$U_n/n! = \sum_1^{n-1} (-1)^{n+1}/(n+1)! + B$$

$$= \sum_2^{n} (-1)^n/n! + B.$$

Hence $\quad U_n = Bn! + n!\left[\frac{1}{2!} - \frac{1}{3!} + \ldots + \frac{(-1)^n}{n!}\right].$

Since $\quad U_2 = 1,$

$$1 = B \,.\, 2! + 2![1/2!],$$

$$B = 0.$$

Thus $\quad U_n = n!\left[\frac{1}{2!} - \frac{1}{3!} + \ldots + \frac{(-1)^n}{n!}\right].$

Note

(*a*) If the parcels are n wires to be fitted into n appropriate plug holes, the probability of no wires being fitted correctly if joined up at random is

$$\frac{U_n}{n!} = \frac{1}{2!} - \frac{1}{3!} + \ldots + \frac{(-1)^n}{n!} \to e^{-1}.$$

The equation could have been solved by writing it

$$U_{n+2} - (n+2)U_{n+1} + U_{n+1} - (n+1)U_n = 0,$$

i.e. if

$$V_n = U_{n+1} - (n+1)U_n$$

$$0 = V_{n+1} + V_n.$$

(*b*) If V_n is the number of ways in which only one is correct then if C is the correct one the remaining $(n-1)$ are incorrect. Since C can be chosen in n ways

$$U_n = nU_{n-1}$$

$$= n!\left[\frac{1}{2!} - \frac{1}{3!} + \ldots + \frac{(-1)^{n-1}}{(n-1)!}\right].$$

Similarly if w_n is the number of ways in which two only are correct

$$w_n = n(n-1)U_{n-2}/2!$$

$$= \frac{n!}{2!}\left[\frac{1}{2!} - \frac{1}{3!} + \ldots + \frac{(-1)^{n-2}}{(n-2)!}\right].$$

12. In testing whether a batch of articles are satisfactory, following Wald and Barnard, we introduce a scoring system for the batch as each sample is examined. The score starts at 0; each time a defective item is encountered we subtract b, while each time the item is effective we add one. As soon as the total score rises to H, the procedure stops and the batch is accepted. As soon as it falls to $-H$, the batch is rejected. This version of Wald's scheme for the qualitative inspection of a set of components, can be compared to a game between two players C and D. Let the skill of the former be measured by p, and that of the latter by q, and the stakes won or lost at each stage be 1 and b respectively, where b is an integer. Each of the players starts the game with H counters, and continues until one or other is bankrupt.

Let $u(x)$ be the probability that D will win when he has x counters. At the next trial he will either win 1 counter from C with probability q, or lose b counters to C with probability p. Thus

$$u(x) = qu(x+1) + pu(x-b).$$

The boundary conditions are that $u(x) = 0$ for $x \leqslant 0$, and $u(2H) = 1$. The equation can be written

$$qu(x+b+1) - u(x+b) + pu(x) = 0.$$

Let $u(x) = q^{-x}v(x)$, then the equation takes the form—

$$v(x + b + 1) - v(x + b) = - cv(x),$$

where $c = pq^b$, and the boundary conditions demand that $v(x) = 0$ for $x \leqslant 0$, and $v(2H) = q^{2H}$. To solve this equation in the orthodox way would involve finding the roots of an equation of the form—

$$z^{b+1} - z^b + c = 0.$$

While these roots can always be obtained by approximation to any required degree of accuracy, in this case this is not necessary. Since $v(x)$ is zero for all negative values of x and for $x = 0$, we shall seek a solution of the difference equation which satisfies the condition $v(2H) = q^{2H}$, accepting it for x positive, and take $v(x) \equiv 0$ for all other values of x. There is no justification for assuming that the same formula must necessarily be applicable both for x positive and for x negative.

Since p and q are positive fractions and b is a positive integer, $c = pq^b$ is less than unity. We seek a solution in $v(x)$ in the form of a power series in c, writing—

$$v(x) = v_0(x) + cv_1(x) + c^2v_2(x) + c^3v_3(x) + \ldots,$$

and insert this in the equation

$$v(x + b + 1) - v(x + b) = - cv(x).$$

On equating the coefficients of the successive powers of c to zero we obtain the following system of equations—

$$\begin{aligned}
v_0(x + b + 1) - v_0(x + b) &= 0,\\
v_1(x + b + 1) - v_1(x + b) &= - v_0(x)\\
v_2(x + b + 1) - v_2(x + b) &= v_1(x)\\
\ldots\ldots\ldots\ldots\ldots&\ldots\ldots\ldots\\
v_r(x + b + 1) - v_r(x + b) &= (-1)^r v_{r-1}(x)\\
\ldots\ldots\ldots\ldots\ldots&\ldots\ldots\ldots
\end{aligned}$$

The first of these leads to a particular form

$$v_0(x) = 1.$$

The second to $v_1(x + b) = - x$,

i.e. $$v_1(x) = - (x - b).$$

The third to $\quad v_2(x + b) = (x - b)(x - b - 1)/2!$

i.e. $\quad v_2(x) = (x - 2b)(x - 2b - 1)/2!$

and so on. Accordingly

$$q^{x}u(x) = v(x). = A[1 - c(x - b) + c^2(x - 2b)(x - 2b - 1)/2!$$
$$- c^3(x - 3b)(x - 3b - 1)(x - 3b - 2)/3! + \ldots].$$

Inserting the boundary condition which determines A, the constant, gives

$$q^{2H} = A[1 - c(2H - b) + c^2(2H - 2b)(2H - 2b - 1)/2! - \text{etc.}].$$

From these expressions it is possible to obtain the exact Operating Characteristic Curve of an inspection scheme with handicap H and penalty b. It is, of course, not our purpose here to examine methods of Quality Control based on these principles. The details of the applications and the theory on which it is based will be found in the Supplement to the *Journal of the Royal Statistical Society*, Vol. VIII, No. 1 (1946), "Sequential Tests in Industrial Statistics" by G. A. Barnard.

EXERCISES (II)

1. In the system of equations $x^2 - p_n x + q_n = 0$, $(n = 1, 2, 3 \ldots)$, the roots of the $(n + 1)$th equation are p_n and $n + 1$; show that

$$p_n = p_1 + (n + 2)(n - 1)/2$$
$$q_n = np_1 + (n - 2)n(n + 1)/2.$$

2. At age A years a man invests £1,000 at $2\frac{1}{2}\%$ and at the end of each year withdraws £x and spends it. The reduced capital, giving him a reduced return in interest is then further reduced by £x at the end of the next year. This process is repeated for N years in succession. At his then age, $A + N$, the man can purchase an annuity of £1 for £p. Show that the value of x, in order that at age $A + N$ he may have sufficient of the capital left to purchase an annuity of £x is given by

$$x = 1000(41/40)^N[p - 40 + 40(41/40)^N].$$

3. A series of simple ovals n in number in a plane are so placed that each oval meets every other oval in two points, No point in the plane is common to more than two ovals. Show that the total number of compartments into which the plane is divided is $n^2 - n + 2$.

4. A series of n spheres are so placed that each intersects all the others. If u_n is the number of compartments into which space is divided by the spheres show that—

$$u_{n+1} = u_n + n^2 - n + 2$$

and that $\quad u_n = n(n^2 - 3n + 8)/3.$

CHAPTER 8

Difference Equations Associated with Functions of Two Variables

Partial Difference Equations with Constant Coefficients

1. We define the partial operators E and F by simple extension of the one-dimensional case as follows

$$\mathrm{E}z(x, y) = z(x + 1, y),$$
$$\mathrm{F}z(x, y) = z(x, y + 1),$$

with difference operators Δ_x and Δ_y such that

$$\mathrm{E} \equiv 1 + \Delta_x \qquad \mathrm{F} \equiv 1 + \Delta_y$$

i.e.
$$\Delta_x z(x, y) = z(x + 1, y) - z(x, y);$$
$$\Delta_y z(x, y) = z(x, y + 1) - z(x, y).$$

Thus $$\mathrm{E}^m\mathrm{F}^n z(x, y) = z(x + m, y + n) = \mathrm{F}^n\mathrm{E}^m z(x, y),$$

and it is easy to establish that the four operators E, F, Δ_x, Δ_y satisfy the rules of ordinary algebra, as a simple extension of the more restricted case with which we have already dealt.

EXERCISE (I)

Show that $\mathrm{E}^r\mathrm{F}^s(\lambda^x\mu^y) = \lambda^r\mu^s \,.\, \lambda^x\mu_z$ and generally that

$$\phi(\mathrm{E}, \mathrm{F})\lambda^x\mu^y = \phi(\lambda, \mu)\lambda^x\mu^y.$$

Accordingly we have the following double interpolation formula, where for this purpose $\mathrm{E}f(a, b) = f(a + h, b)$, $\mathrm{F}f(a, b) = f(a, b + k)$,

$$\begin{aligned} f(a + xh, b + yk) &= \mathrm{E}^x\mathrm{F}^y f(a, b) = (1 + \Delta_x)^x(1 + \Delta_y)^y f(a, b) \\ &= \{1 + x\Delta_x + y\Delta_y \\ &\qquad + [x(x - 1)\Delta_x^2 + 2xy\Delta_x\Delta_y + y(y - 1)\Delta_y^2]/2! \\ &\qquad + \ldots\}f(a, b). \end{aligned}$$

Also, corresponding approximate formulae for $\dfrac{\partial u}{\partial x}, \dfrac{\partial u}{\partial y}, \dfrac{\partial^2 u}{\partial x\,\partial y}$, viz.

$$\mathrm{E} \equiv e^{h\frac{\partial}{\partial x}} \quad \text{and} \quad \mathrm{F} \equiv e^{k\frac{\partial}{\partial y}}$$

or

$$h\frac{\partial}{\partial x} \equiv \log(1 + \Delta_x), \quad k\frac{\partial}{\partial y} \equiv \log(1 + \Delta_y).$$

EXERCISES (II)

1. Show that $\Delta_x\Delta_y z(x, y) \equiv [\mathrm{EF} - \mathrm{E} - \mathrm{F} + 1]z(x, y)$.
2. Show that $u(x + 1, y + 1) + u(x + 1, y - 1) + u(x - 1, y + 1) + u(x - 1, y - 1) - 4u(x, y)$
$\equiv [\mathrm{E}(\mathrm{F} + i) - i(\mathrm{F} - i)][\mathrm{E}(\mathrm{F} - i) + i(\mathrm{F} + i)]u(x - 1, y - 1)$.
3. Using the identities $\Delta_x{}^m\Delta_y{}^n u(x, y) \equiv (\mathrm{E} - 1)^m(\mathrm{F} - 1)^n u(x, y)$, express $\Delta_x{}^m\Delta_y{}^n u(x, y)$ in terms of the values of u at (x, y), $(x + 1, y)$, $(x, y + 1)$, $(x, y + 2)$, etc.
4. Express $u(x + m, y + n)$ in terms of the differences of $u(x, y)$.

A partial difference equation is a given functional relation between

$$u(x, y),\ u(x + 1, y),\ u(x, y + 1),\ u(x + 1, y + 1),\ u(x + 2, y), \text{ etc.}$$

that is, between

$$u(x, y),\ \mathrm{E}u(x, y),\ \mathrm{F}u(x, y),\ \mathrm{EF}u(x, y),\ \mathrm{E}^2u(x, y), \text{ etc.}$$

Thus, for example,

$$u(x + 2, y + 1) = u(x, y)^2 - 2u(x + 1, y)u(x, y + 1)$$

is a partial difference equation.

If $x + r$ and $y + s$ are the terms involving the largest positive values of r and s respectively in an equation which contains $u(x, y)$, the equation is said to be of the rth order in x and the sth order in y. The distinction is important in respect to the boundary conditions that require to be associated with the partial difference equation to express the function uniquely everywhere.

Thus for instance if, in the above example, $u(x, y)$ is uniquely specified at every point (x, y) in the region $0 < x < 2$, $-\infty < y < \infty$, then by using the expression for $u(x + 2, y + 1)$ above, u may be evaluated step-by-step at every point (x, y) in the plane x, y. Again if $u(x, y)$ is uniquely specified at every point in the region

$-\infty < x < \infty, 0 \leqslant y < 1$, the same equation again enables $u(x, y)$ to be evaluated at every finite point (x, y) in the plane. The one set of boundary conditions relates to an infinite band parallel to the axis of y, of width 2, the other to an infinite band of width 1, parallel to the axis of x.

We shall see shortly that the explicit solution of a partial difference equation involves certain arbitrary functions of the variables x and y, so that the final determination of the solution that fits given boundary conditions requires the precise determination of these functions. Since any difference equation is in fact also a functional equation, the process of solution is in reality the conversion of one functional equation into another.

Consider the expression

$$z(x, y) = f[x, y, a, \phi(a)], \tag{1}$$

where f is a given function of the arbitrary constant a and $\phi(a)$ an arbitrary function of a.

There is one and only one difference equation of the type

$$F[x, y, z(x, y), z(x + 1, y), z(x, y + 1)] = 0, \tag{2}$$

satisfied by the above expression for z. For

$$z(x + 1, y) = f[x + 1, y, a\,\phi(a)], \tag{3}$$

$$z(x, y + 1) = f[x, y + 1, a, \phi(a)]. \tag{4}$$

By eliminating a and $\phi(a)$ between equations (1), (3), and (4) a unique expression of the form (2) is obtained.

We propose to present (1) which is a very general form of the solution of (2), since a and $\phi(a)$ are arbitrary, in a different form.

In (1) let $z(0, 0)$ be the value of z at $x = 0$ $y = 0$ then

$$z(0, 0) = f[0, 0, a, \phi(a)]. \tag{5}$$

Now let (1) and (5) be solved for a and $\phi(a)$ giving

$$a = \psi[x, y, z, z(0, 0)]$$
$$\phi(a) = \chi[x, y, z, z(0, 0)].$$

Hence a form equivalent to (1) but involving the value of z at $(0, 0)$ is

$$\chi[x, y, z,(0, 0)] = \phi[\psi(x, y, z, z(0, 0))], \tag{6}$$

where χ and ψ are definite determinate functions of x, y, z, $z(0, 0)$ and ϕ is arbitrary. Thus the solution (6) of (2) here presents itself in an equivalent form involving one arbitrary function.

The foregoing procedure passes from the solution (1) to the eliminant difference equation (2). Since f in (1) may be any function, equation (2) may be any equation provided it is of the general form indicated. It follows therefore that the solution of a difference equation of type (2) involves one arbitrary function of the variables.

2. Every linear partial difference equation with constant coefficients can be written in the form

$$\phi(\mathrm{E}, \mathrm{F})u(x, y) = R(x, y),$$

where the operator $\phi(\mathrm{E}, \mathrm{F})$ is a finite polynomial function both in E and F, and $R(x, y)$ is a given function of x and y.

We commence with the case where $R(x, y)$ is absent, reserving the determination of the particular solution for later examination. Consider therefore,

$$\phi(\mathrm{E}, \mathrm{F})u(x, y) = 0.$$

In general we may suppose $\phi(\mathrm{E}, \mathrm{F})$ factorizable in the form

$$\phi(\mathrm{E}, \mathrm{F}) = \prod [\mathrm{E} - \psi_r(\mathrm{F})]^r,$$

where $\psi_r(\mathrm{F})$ may be an irrational function of F, and the sum of the powers r is the degree of E in the function $\phi(\mathrm{E}, \mathrm{F})$.

We restrict ourselves in the first instance to the simple linear case

$$(\mathrm{E} - a\mathrm{F} - b)u(x, y) = 0,$$

i.e. $$u(x + 1, y) = au(x, y + 1) + bu(x, y).$$

As far as its effect on x is concerned $a\mathrm{F} + b$ acts as a constant, so that formally we may write

$$u(x, y) = (a\mathrm{F} + b)^x A,$$

where A, a unit periodic in x, is arbitrary as regards y. Thus

$$\begin{aligned} u(x, y) &= (a\mathrm{F} + b)^x A(y) \\ &= b^x \left(1 + \frac{a\mathrm{F}}{b}\right)^x A(y) \\ &= b^x \left(1 + \frac{ax\mathrm{F}}{b} + \frac{a^2x(x-1)\mathrm{F}^2}{b^2 2!} + \dots\right) A(y) \end{aligned}$$

$$= b^x \left[A(y) + \frac{ax}{b} A(y+1) + \frac{a^2 x(x-1)}{b^2 2!} A(y+2) + \ldots \right]$$

where as we have stated A is also a unit periodic in x. If x is a positive integer this series terminates.

We notice therefore that the linear partial difference equation

$$u(x+1, y) - au(x, y+1) - bu(x, y) = 0$$

has in its solution one arbitrary function of y, viz. $A(y)$. An alternative method of presenting the solution could be derived as follows. Write the equation in the form

$$\left(\mathrm{F} - \frac{1}{a} (\mathrm{E} - b)\, u(x, y) \right) = 0$$

so that following a similar procedure

$$u(x, y) = \frac{1}{a^y} (\mathrm{E} - b)^y B(x)$$

$$= \left(-\frac{b}{a} \right)^y \left(1 - \frac{\mathrm{E}}{b} \right)^y B(x)$$

$$= \left(-\frac{b}{a} \right)^y \left[B(x) - \frac{y}{b} B(x+1) + \frac{y(y-1)}{b^2 2!} B(x+2) + \ldots \right],$$

again in terms of a single arbitrary function.

The one form of solution, as we shall shortly see, would be appropriate for boundary conditions that relate to a band parallel to the axis of y, the other to a band parallel to the axis of x.

Since in the factorization of $\phi(\mathrm{E}, \mathrm{F})$ regarded as a polynomial in either E or F, repeated factors may occur we proceed to examine this case.

Multiple Linear Factors

Suppose

$$u(x+2, y) + a^2 u(x, y+2) + b^2 u(x, y) - 2au(x+1, y+1) - 2bu(x+1, y) + 2abu(x, y+1) = 0,$$

i.e.
$$(\mathrm{E} - a\mathrm{F} - b)^2 u(x, y) = 0.$$

Thus

$$u(x, y) = (a\text{F} + b)^x[A(y) + xB(y)]$$

$$= b^x\left(1 + \frac{a}{b}\text{F}\right)^x[A(y) + xB(y)]$$

$$= b^x\left[A(y) + \frac{ax}{b}A(y+1) + \frac{a^2}{b^2}\frac{x(x-1)}{2!}A(y+2) + \ldots\right]$$

$$+ \left[xB(y) + \frac{ax^2}{b}B(y+1) + \frac{a^2x^2(x-1)}{b^2 2!}B(y+2) + \ldots\right]$$

$$= b^x\left[[A(y) + xB(y)] + \frac{ax}{b}[A(y+1) + xB(y+1)]\right.$$

$$\left. + \frac{a^2(x-1)}{b^2 2!}[A(y+2) + xB(y+2)] + \ldots\right].$$

EXERCISE (III)

Solve $u(x+2, y) + u(x, y+2) + u(x, y)$

$$= 2[u(x+1, y+1) - u(x+1, y) + u(x, y+1)].$$

More generally, consider $(\text{E} - a\text{F} - b)^r u(x, y) = 0$.

Since the solution of $$(\text{E} - \alpha)^r y(x) = 0$$

is $$y(x) = \alpha^n\left[A_1 + xA_2 + \frac{x(x-1)}{2!}A_3\right.$$

$$\left. + \ldots + \frac{x(x-1)\ldots(x-r+1)}{r!}A_r\right]$$

the solution here is clearly

$$u(x, y) = (a\text{F} + b)^x\left[A_1(y) + xA_2(y) + \frac{x(x-1)}{2!}A_3(y) + \ldots\right]$$

$$= b^x\left(1 + \frac{a\text{F}}{b}\right)^x\left[A_1(y) + xA_2(y) + \frac{x(x-1)}{2!}A_3(y) + \ldots\right],$$

and by expanding $\left(1 + \frac{a}{b}\text{F}\right)^x$, in ascending powers of F and operating on the r functions $A_1(y)$, $A_2(y)$, . . ., $A_r(y)$ the general form of the solution is at once obtainable.

The difference equation is clearly of the rth order in x and the solution found contains r arbitrary and independent functions of y.

EXERCISE (IV)

What is the alternative form in terms of arbitrary functions of x?

If in the equation $\quad \phi(E, F)u(x, y) = 0$

the operator $\phi(\mathrm{E}, \mathrm{F})$ can be written in the form

$$\phi(\mathrm{E}, \mathrm{F}) = (\mathrm{E} + a_1\mathrm{F} + b_1)^{r_1}(\mathrm{E} + a_2\mathrm{F} + b_2)^{r_2} \ldots (\mathrm{E} + a_n\mathrm{F} + b_n)^{r_n},$$

then since solutions are additive and the separate factors of $\phi(\mathrm{E}, \mathrm{F})$ are commutative, the general solution of the equation will be the sum of the general solutions of the separate equations

$$(\mathrm{E} + a\mathrm{F} + b)^r u(x, y) = 0.$$

Example

$$u(x + 2, y) - u(x, y + 2) - 2u(x + 1, y) + 2u(x, y + 1) = 0$$

Here $\qquad (\mathrm{E} - \mathrm{F})(\mathrm{E} - 2 - \mathrm{F})u(x, y) = 0.$

Since the solution of the equation $(\mathrm{E} - \mathrm{F})u(x, y)$ is

$$\mathrm{F}^x B(y) = B(x + y)$$

where B is an arbitrary function of its variable,

$$\begin{aligned} u(x, y) &= B(x + y) + 2^x(1 - \tfrac{1}{2}\mathrm{F})^x A(y) \\ &= B(x + y) + 2^x(A(y) + \tfrac{1}{2}xA(y + 1) + \ldots). \end{aligned}$$

To specify this solution exactly it suffices to know $u(x, y)$ over either of the bands

$$0 \leqslant x < 2, \qquad -\infty < y < +\infty$$

or $\qquad 0 \leqslant y < 2, \qquad -\infty < x < +\infty.$

When the factors are not linear but expressible as regards E or F in terms of simple irrationals a corresponding procedure may be adopted. For example

$$u(x + 2, y) - 4u(x, y + 1) = 0,$$

gives $\qquad (\mathrm{E}^2 - 4\mathrm{F})u(x, y) = 0$

or $\qquad (\mathrm{E} - 2\mathrm{F}^{\frac{1}{2}})(\mathrm{E} + 2\mathrm{F}^{\frac{1}{2}})u(x, y) = 0$

Thus $\qquad u(x, y) = 2^x[\mathrm{F}^{x/2}A(y) + (-1)^x\mathrm{F}^{x/2}B(y)]$

$$= 2^x A(y + x/2) + (-2)^x B(y + x/2).$$

Alternatively $\qquad u(x, y) = 4^{-y}C(x + 2y).$

EXERCISE (V)

Solve the equation $u(x + 2, y) = u(x, y + 1) + u(x, y)$, by writing the operator both as

$$(E + \sqrt{1 + F})(E - \sqrt{1 + F}) \quad \text{and} \quad (F + 1 - E^2),$$

and identify the two forms of solutions obtained.

Lagrange's Method of Solution

Consider
$$\phi(E, F)u(x, y) = 0,$$

where $\phi(E, F)$ is a polynomial function in E and F.

We seek a solution of the form $\lambda^x\mu^y$ in the first instance. By inserting in the equation it follows immediately that this will be satisfied if

$$\phi(\lambda, \mu) = 0.$$

With this relation between λ and μ, any arbitrary constant may be attached to the term $\lambda^x\mu^y$, and since the equation is linear the sum of all such terms with all possible values of λ and μ will also be a solution. Thus

$$u(x, y) = \sum_r A(\lambda_r, \mu_r)\lambda_r^x\mu_r^y,$$

is a solution, or alternately if we regard λ as varying from $-\infty$ to $+\infty$

$$u(x, y) = \int_{-\infty}^{\infty} A(\lambda, \mu)\lambda^x\mu^y d\lambda$$

where
$$\phi(\lambda, \mu) = 0.$$

Examples

1.
$$u(x + 1, y + 1) = au(x, y).$$
$$(EF - a)u(x, y) = 0.$$

This demands the relation

$$\lambda\mu - a = 0$$

and the solution takes the form

$$\Sigma A(\lambda, a/\lambda)\lambda^x a^y \lambda^{-y} = \Sigma A(\lambda, a/\lambda)a^y\lambda^{x-y}$$
$$= a^y\Sigma C_\lambda\lambda^{x-y},$$

where C_λ is a set of arbitrary constants. This means in effect that the derived solution is

$$u(x, y) = a^y B(x - y)$$

where B is arbitrary.

It is easily verified that this is the form that would have been derived by writing the equation

$$(\mathrm{F} - a\mathrm{E}^{-1})u(x, y) = 0$$

and solving as before. Equally, of course, it could be written

$$u(x, y) = a^x C(x - y).$$

2. A and B engage in transactions with each other. The result of each transaction is that a coin passes from one to the other. The probability of a transaction being to A's advantage is p. Each proposes to retire when and if he acquires N coins. What is the probability of A retiring?

Let x and y be the number of coins A and B possess at any stage. Then x and y are positive integers and $x + y = k$, a constant. Let $u(x, y)$ be the probability of A finally succeeding, as estimated at the x, y stage. The probability of A finally retiring is the probability of A gaining a coin at the next step and thereafter finally winning the necessary number N, plus the probability of A losing the next coin and thereafter finally winning the necessary number.

Now if A wins a coin at the (x, y) stage the probability of his winning N finally is $u(x + 1, y - 1)$, while if A loses a coin at the (x, y) stage the probability of his finally acquiring N is $u(x - 1, y + 1)$. Hence

$$u(x, y) = pu(x + 1, y - 1) + (1 - p)u(x - 1, y + 1).$$

This is the partial difference equation that describes the process of change in the probability of A finally being able to retire. Moreover, when A has N coins (at which stage B has $k - N$), the probability of A retiring is 1. Hence

$$u(N, k - N) = 1.$$

In the same way since B retires when he has N coins

$$u(k - N, N) = 0.$$

These are the boundary conditions that $u(x, y)$ has to satisfy. The general solution of the difference equation is

$$u(x, y) = \left(\frac{1}{p} - 1\right)^x \phi(x + y) + \psi(x + y)$$

where ϕ and ψ are arbitrary functions.

Inserting the boundary conditions, we find

$$1 = \left(\frac{1}{p} - 1\right)^N \phi(k) + \psi(k) = q^N\phi(k) + \psi(k),$$

$$0 = \left(\frac{1}{p} - 1\right)^{k-N} \phi(k) + \psi(k) = q^{k-N}\phi(k) + \psi(k),$$

where $q = \dfrac{1}{p} - 1.$

Hence
$$\phi(k) = q^N/(q^{2N} - q^k)$$
$$\psi(k) = q^k/(q^k - q^{2N}).$$

Inserting these forms for ϕ and ψ in the expression for $u(x, y)$ we have finally—

$$\begin{aligned} u(x, y) &= q^x\phi(x + y) + \psi(x + y) \\ &= q^{x+N}/(q^{2N} - q^{x+y}) + q^{x+y}/(q^{x+y} - q^{2N}) \\ &= (q^{N-y} - 1)/(q^{2N-x-y} - 1) \\ &= p^{N-x}[(1 - p)^{N-y} - p^{N-y}]/[(1 - p)^{2N-x+y} - p^{2N-x+y}]. \end{aligned}$$

EXERCISES (VI)

1. (i) Verify that if $x = y$: $p = \frac{1}{2}$, $u = \frac{1}{2}$; $p = 1$, $u = 1$; $p = 0$, $u = 0$.

(ii) Show that if the probability of A winning the coin is 1/3 and if at any stage A and B both have the same number of coins, the probability of B retiring is $1/[1 + 2^{-N/2}]$.

2. If $\phi(\mathrm{E}, \mathrm{F})$ is a polynomial function of the single product (EF), solve the equation $\phi(\mathrm{EF})u(x, y) = 0$, examining particularly the case of multiple factors of $\phi(\mathrm{EF})$.

3. A region in the x, y plane in which a function u satisfies the equation $\dfrac{\partial^2 u}{\partial x^2} + \dfrac{\partial^2 u}{\partial y^2} = 0$ is divided into equal small squares of side ω by lines parallel to x and y.

Show that if $v(x + r, y + s) = v(x + r\omega, y + s\omega)$, v approximately satisfies the equation—

$$v(x + 1, y + 1) + v(x + 1, y + 1) + v(x - 1, y + 1) + v(x - 1, y - 1) = 4v(x, y),$$

and specify the degree of approximation.

Show further that this difference equation has a solution of the form

$$u(x, y) = \Sigma C \cot^y (\pi/4 - d) \sin (2x\alpha + B),$$

where the summation extends over values of the arbitrary constants B, C, α.

The Particular Solution

The equation is

$$\phi(\mathrm{E}, \mathrm{F})u(x, y) = R(x, y).$$

We consider some special forms.

(*a*) $R(x, y)$ is a polynomial function of x and y of finite degree.

Write
$$\phi(\mathrm{E}, \mathrm{F}) = \phi(1 + \Delta_x, 1 + \Delta_y),$$

and expand $1/\phi(1 + \Delta_x, 1 + \Delta_y)$, in ascending powers of Δ_x and Δ_y. Accordingly we obtain the particular solution in the form—

$$u(x, y) = [A_0 + A_1\Delta_x + B_1\Delta_y + A_2\Delta_x^2 + B_2\Delta_y^2 + C_2\Delta_x\Delta_y + \ldots]R(x, y),$$

and these operations, when carried through, lead to a finite number of terms.

Example

$$u(x + 1, y + 1) + 2u(x, y) = x^2 + y + 1.$$

Thus
$$(\mathrm{EF} + 2)u(x, y) = x^2 + y + 1.$$

Accordingly

$$(\mathrm{EF} + 2)^{-1} = \tfrac{1}{3}[1 + \tfrac{1}{3}(\Delta_x + \Delta_y + \Delta_x\Delta_y)]^{-1}$$
$$= \tfrac{1}{3}[1 - \tfrac{1}{3}(\Delta_x + \Delta_y + \Delta_x\Delta_y) + \tfrac{1}{9}(\Delta_x^2 + 2\Delta_x\Delta_y + \ldots) \ldots],$$

and this when operating on $x^2 + y + 1$ gives

$$\tfrac{1}{3}[x^2 + y + 1 - \tfrac{1}{3}(2x + 1 + 1) + \tfrac{1}{9}\,.\,2] = x^2/3 - 2x/9 + y/3 + 5/27$$

which is easily verified to be a particular solution.

For the complementary function in this case we have that

$$(\mathrm{EF} + 2)u(x, y) = 0$$

is equivalent to $$(\mathrm{E} + 2\mathrm{F}^{-1})u(x, y) = 0$$

or $$u(x, y) = (-2)^x \mathrm{F}^{-x} \,.\, A(y)$$

$$= (-2)^x A(y - x).$$

Thus the complete solution is immediately obtained by adding the complementary function to the particular solution.

EXERCISE (VII)

If $(a^{-m}\mathrm{E})(a^{-m}\mathrm{E}) \ldots n$ times $= a^{u(m,\, n)}\mathrm{E}^n$, show that

$$u(m + 1, n) - u(n, m + 1) = m.$$

Hence prove that $(a^{-m}\mathrm{E})^n = a^{-n(2m+n-1)/2}\mathrm{E}^n$.

(*b*) If $$R(x, y) = \lambda^{ax+by} S(x, y)$$

where λ, a, and b are constants and $S(x, y)$ is a polynomial of finite degree in x and y, we begin by writing

$$u(x, y) = \lambda^{ax+by} v(x, y).$$

It is easily shown that

$$\mathrm{E}^r\mathrm{F}^s \lambda^{ax+by} v(x, y) = \lambda^{ar+bs} \lambda^{ax+by} \mathrm{E}^r\mathrm{F}^s v(x, y).$$

Thus $$\phi(\mathrm{E}, \mathrm{F})\lambda^{ax+by} v(x, y) = \lambda^{ax+by} \phi(\lambda^a\mathrm{E}, \lambda^b\mathrm{F}) v(x, y),$$

by this transformation the equation reduces, by removing λ^{ax+by} from both sides, to the case (*a*) already dealt with.

(*c*) *Examples of a Special Kind*

Example 1

$$u(x + 1, y + 1) - u(x, y) = P(x - y).$$

Here we notice that the function on the right-hand side is of the same form as the complementary function $A(x - y)$. Write

$$u(x, y) = P(x - y) v(x, y),$$

then the equation becomes

$$v(x + 1, y + 1) - v(x, y) = 1$$

so that $$v(x, y) = A(x - y) + \tfrac{1}{2}(x + y)$$

and $$u(x, y) = P(x - y)A(x - y) + \tfrac{1}{2}(x + y)P(x - y)$$

$$= B(x - y) + \tfrac{1}{2}(x + y)P(x - y),$$

where $B(x - y)$ is an arbitrary function of $x - y$.

Example 2

$$u(x+1, y+1) - u(x, y) = a^x P(x-y)\chi(x, y)$$

where P is any given function of $(x - y)$, and $\chi(x, y)$ is any given polynomial in x and y.

Write $$u(x, y) = a^x P(x-y)v(x, y),$$

then $$u(x+1, y+1) = a \,.\, a^x P(x-y)v(x+1, y+1)$$

and the equation reduces to

$$v(x+1, y+1) - v(x, y) = \frac{1}{a}\chi(x, y)$$

and this is immediately soluble by the previous methods.

Example 3

$$u(x+1, y) - au(x, y+1) = a^x f(x+y)\chi(x, y)$$

where $\chi(x, y)$ is a polynomial function in x and y. Write

$$u(x, y) = a^x f(x+y-1)v(x, y),$$

then $$v(x+1, y) - v(x, y+1) = \frac{1}{a}\chi(x, y)$$

and the equation is immediately soluble.

EXERCISES (VIII)

Solve the following difference equations—

1. $u(x+1, y) - 2u(x, y+1) = (x+y)\, e^{x-2y}$.
2. $u(x+2, y) - 2u(x+1, y+1) + u(x, y+2) = x + y$.
3. $u(x+2, y) - 3u(x+1, y+1) + 2u(x, y+2) = y \,.\, 2^x$.
4. $u(x+2, y+2) - 9u(x+1, y+1) + 20u(x, y) = xy$.
5. $u(x+1, y+1) - u(x, y) = (x-y)^2$.
6. $u(x+1, y+1) - 2u(x, y) = 2^x(x^2 - y^2)$.

Simultaneous System of Linear Partial Difference Equations with Constant Coefficients

We consider the case of two functions u and v of two independent variables x and y. The two functions are connected through two linear relations. For example

$$4u(x+1, y+1) - 3v(x+1, y+1) - 2u(x, y) + v(x, y) = R(x, y),$$
$$2u(x+1, y+1) + v(x+1, y+1) - 2u(x, y) - v(x, y) = S(x, y),$$

constitute a simultaneous pair to determine $u(x, y)$ and $v(x, y)$, when $R(x, y)$ and $S(x, y)$ are given.

In common with all such systems the general solution will consist of the complementary function which will contain the arbitrary functions that arise, and the particular solution which will allow for the existence of $R(x, y)$ and $S(x, y)$ on the right-hand side.

If E and F represent two operators acting respectively on x and y in the usual way, then the illustration above may be written

$$2(2\mathrm{EF} - 1)u(x, y) - (3\mathrm{EF} - 1)v(x, y) = R(x, y),$$
$$2(\mathrm{EF} - 1)u(x, y) + (\mathrm{EF} - 1)v(x, y) = S(x, y).$$

In general a simultaneous pair of partial difference equations can be written

$$f_1(\mathrm{E}, \mathrm{F})u(x, y) + g_1(\mathrm{E}, \mathrm{F})v(x, y) = R(x, y),$$
$$f_2(\mathrm{E}, \mathrm{F})u(x, y) + g_2(\mathrm{E}, \mathrm{F})v(x, y) = S(x, y).$$

The method of solution then proceeds operationally in parallel with that adopted for the case of one independent variable, and analogous with that for differential equations, care being taken to ensure by substitution in the equations that the proper number of arbitrary functions are present in the final form of solution. We illustrate with the above example where $R(x, y) = 1$ and $S(x, y) = x$.

Example

$$2(2\mathrm{EF} - 1)u(x, y) - (3\mathrm{EF} - 1)v(x, y) = 1$$
$$2(EF - 1)u(x, y) + (\mathrm{EF} - 1)v(x, y) = x.$$

We note that in this case it would be possible to solve the second equation at once for $2u(x, y) + v(x, y)$ and then utilize this result in the first equation directly, but we shall not do so here since it is special to this equation and would not illustrate the general method. Accordingly we operate on the first equation with (EF − 1) and on the second with (3EF − 1) and add. This eliminates $v(x, y)$. Thus

$$[2(2\mathrm{EF} - 1)(\mathrm{EF} - 1) + 2(\mathrm{EF} - 1)(3\mathrm{EF} - 1)]u(x, y)$$
$$= (\mathrm{EF} - 1) \,.\, 1 + (3\mathrm{EF} - 1)x$$
$$= 1 - 1 + 3(x + 1) - x = 2x + 3,$$

so that $$(\mathrm{EF} - 1)(5\mathrm{EF} - 2)u(x, y) = x + 3/2.$$

Hence $$u(x, y) = f(x - y) + (\tfrac{2}{5})^x g(x - y) + \frac{x(3x - 4)}{18}.$$

Again eliminating $u(x, y)$ in the original pair gives

$$(\mathrm{EF} - 1)(5\mathrm{EF} - 2)v(x, y) = 2x + 1$$

so that $$v(x, y) = F(x - y) + (\tfrac{2}{5})^x G(x - y) + \frac{x(3x - 1)}{18}.$$

On inserting these expressions for $u(x, y)$ and $v(x, y)$ in the original pair we have finally—

$$u(x, y) = f(x - y) + (\tfrac{2}{5})^x g(x - y) + x(3x - 4)/18,$$
$$v(x, y) = f(x - y) - 2(\tfrac{2}{5})^x g(x - y) + x(3x - 1)/18,$$

where the functions f and g are arbitrary.

Some Special Forms of Partial Difference Equations

(*a*) Certain simple transformations of the dependent variable, similar to those applied to ordinary difference equations, can be used on occasion to reduce a non-linear equation to linear form. For example the equation

$$z(x + 1, y + 1) = z(x + 1, y)^p z(x, y)^q$$

becomes $$0 = w(x + 1, y + 1) - pw(x + 1, y) - qw(x, y)$$

by writing $$w(x, y) = \log z(x, y).$$

The equation

$$0 = [z(x + 1, y)]^2 - a[z(x + 1, y + 1)]^2 + b[z(x, y)]^2$$

becomes of linear form by writing

$$w(x, y) = [z(x, y)]^2.$$

EXERCISE (IX)

Solve $z(x + 1, y) = z(x, y + 1)^p$ where p is a constant.

(*b*) *Clairaut's Extended Form.* If A and B are arbitrary unit periodics in x and y then the expression

$$z(x, y) = Ax + By + f(A, B)$$

is a solution of the equation

$$z(x, y) = x\Delta_x z(x, y) + y\Delta_y z(x, y) + f[\Delta_x z(x, y), \Delta_y z(x, y)].$$

EXERCISE (X)

Solve—

$$z(x, y)[1 + x + y + z(x, y + 1) + 2(x + 1, y) - z(x, y)]$$
$$= xz(x + 1, y) + yz(x, y + 1) + z(x + 1, y) \,.\, z(x, y + 1).$$

(c) The equation

$$u(m + 1, n + 1) - \phi(m, n)u(m, n) = \psi(m, n)$$

can be solved in finite terms if any particular solution $u(m, n)$ is known for the equation

$$u_1(m + 1, n + 1) - \phi(m, n)u_1(m, n) = 0.$$

For eliminating $\phi(m, n)$ between the two expressions above

$$\frac{u(m + 1, n + 1)}{u_1(m + 1, n + 1)} - \frac{u(m, n)}{u_1(m, n)} = \frac{\psi(m, n)}{u_1(m + 1, n + 1)},$$

or $$v(m + 1, n + 1) - v(m, n) = \chi(m, n)$$

where $v(m, n) = \dfrac{u(m, n)}{u_1(m, n)}, \qquad \chi(m, n) = \dfrac{\psi(m, n)}{u_1(m + 1, n + 1)}.$

Hence

$$v(m + 1, m + n + 1) - v(m, m + n) = \chi(m, m + n),$$

i.e. $$\Delta_m v(m, m + n) = \chi(m, m + n).$$

Thus $$v(m, m + n) = \sum_{1}^{m-1} \chi(m, m + n) + A(n)$$

or $$v(m, m + n) = A(n) + \chi(1, n + 1) + \chi(2, n + 2) + \ldots + \chi(m - 1, m + n - 1),$$

i.e. $$v(m, n) = A(n - m) + \chi(1, n - m + 1) + \chi(2, n - m + 2) + \ldots + \chi(m - 1, n - 1),$$

and $u(m, n)$ follows immediately.

EXERCISES (XI)

1. Show that the solution of the equation—

$$u(m + 1, n + 1) - u(m, n) = \psi(m, n)$$

can be presented in the form

$$u(m, n) = A(m - n) + \psi(1, n - m + 1) + \psi(2, n - m + 2) + \ldots + \psi(m - 1, n - 1).$$

2. Using the foregoing result and writing $v(m, n) = \log u(m, n)$ show that a particular solution of the equation

$$u(m+1, n+1) - \phi(m, n)u(m, n) = 0$$

can always be found.

(*d*) Consider the equation

$$R(m, n) = P(m)z(m+1, n) + Q(n)z(m, n+1) - z(m, n)$$

We write $$u(m, n) = z(m, n) \prod_a^{m-1} P(m) \prod_b^{n-1} Q(n)$$

and, on multiplying by the two product terms, the equation reduces to

$$R(m, n) \prod_a^{n-1} P(m) \prod_b^{n-1} Q(n) = u(m+1, n) + u(m, n+1) - u(m, n)$$

which has a solution in finite terms provided the function $R(m, n)$ is either a polynomial or an exponential function of m and n.

Example

$$m^2 z(m+1, n) + \alpha^n z(m, n+1) - z(m, n) = 0.$$

Writing

$$\begin{aligned} u(m, n) &= 1^2 2^2 3^2 \ldots (m-1)^2 \alpha^0 \alpha^1 \alpha^2 \ldots \alpha^{n-1} z(m, n) \\ &= [(m-1)!]^2 \alpha^{n(n-1)/2} z(m, n), \end{aligned}$$

we have

$$0 = \frac{u(m+1, n)}{(m-1)!^2 \alpha^{n(n-1)/2}} + \frac{u(m, n+1)}{(m-1)!^2 \alpha^{n(n-1)/2}} - \frac{u(m, n)}{(m-1)!^2 \alpha^{n(n-1)/2}},$$

i.e. $$0 = (\mathbf{E} + \mathbf{F} - 1)u(m, n).$$

Hence
$$\begin{aligned} u(m, n) &= (1 - \mathbf{F})^m \phi(n) \\ &= \phi(n) - m\phi(n+1) + \frac{m(m-1)}{2!} \phi(n+2) \\ &\qquad + \ldots + (-1)^m \phi(m+n). \end{aligned}$$

(*e*) Suppose

$$u(x+1, y+1) = \frac{au(x, y) + bv(x, y) + c}{pu(x, y) + qv(x, y) + r}$$

$$v(x+1, y+1) = \frac{Au(x, y) + Bv(x, y) + C}{pu(x, y) + qv(x, y) + r}$$

where a, b, c, A, B, C, p, q, r are constants. From this it follows that if λ, μ and ν are three undetermined multipliers,

$$\frac{\lambda u(x+1, y+1) + \mu v(x+1, y+1) + \nu}{(a\lambda + A\mu + p\nu)u(x, y) + (b\lambda + B\mu + q\nu)v(x, y) + (c\lambda + C\mu + r\nu)}$$

$$= \frac{u(x+1, y+1)}{au(x, y) + bv(x, y) + c} = \frac{v(x+1, y+1)}{Au(x, y) + Bv(x, y) + C}$$

$$= \frac{1}{pu(x, y) + qv(x, y) + r}.$$

Now suppose λ, μ, and ν are chosen to make—

$$\begin{aligned} a\lambda + A\mu + p\nu &= k\lambda \\ b\lambda + B\mu + q\nu &= k\mu \\ c\lambda + C\mu + r\nu &= k\nu \end{aligned}$$

where k is some constant. If λ, μ, and ν are not to be zero this demands that

$$\begin{vmatrix} a-k & A & p \\ b & B-k & q \\ c & C & r-k \end{vmatrix} = 0.$$

This being a cubic equation in k provides three values, k_1, k_2, and k_3 and for each of these we can find the corresponding values of $\lambda : \mu : \nu$.

This enables us therefore to replace the original set of equations by

$$\frac{P_1(x+1, y+1)}{k_1 P_1(x, y)} = \frac{P_2(x+1, y+1)}{k_2 P_2(x, y)} = \frac{P_3(x+1, y+1)}{k_3 P_3(x, y)}$$

where

$$\begin{aligned} P_1 &= \lambda_1 u(x, y) + \mu_1 v(x, y) + \nu_1 \\ P_2 &= \lambda_2(x, y) + \mu_2 v(x, y) + \nu_2 \\ P_3 &= \lambda_3 u(x, y) + \mu_3 v(x, y) + \nu_3. \end{aligned}$$

These are equivalent to

$$\frac{P_1(x+1, y+1)}{P_3(x+1, y+1)} = \frac{k_1}{k_3}\frac{P_1(x, y)}{P_3(x, y)}, \quad \frac{P_2(x+1, y+1)}{P_2(x, y)} = \frac{k_2}{k_3}\frac{P_2(x, y)}{P_3(x, y)}.$$

Equations of this type, viz.

$$f(x+1, y+1) = kf(x, y),$$

have already been solved. Thus

$$\frac{P_1(x, y)}{P_3(x, y)} = \left(\frac{k_1}{k_3}\right)^x \phi_1(x - y) = \left(\frac{k_1}{k_3}\right)^x \frac{\psi_1(x - y)}{\psi_3(x - y)},$$

say
$$\frac{P_2(x, y)}{P_3(x, y)} = \left(\frac{k_2}{k_3}\right)^x \frac{\psi_2(x - y)}{\psi_3(x - y)}$$

or

$$P_1(x, y) : P_2(x, y) : P_3(x, y) = k_1^x \psi_1(x - y) : k_2^x \psi_2(x - y) : k_3^x \psi_3(x - y).$$

Hence by using the expressions for P_1, P_2, and P_3 in terms of λ, μ, ν, $u(x, y)$ and $v(x, y)$, the expressions for the latter are easily found.

Example

$$\frac{u(x + 1, y + 1)}{v(x, y) + 2} = \frac{v(x + 1, y + 1)}{u(x, y) + 1} = \frac{1}{u(x, y) - v(x, y)}$$

$$= \frac{u(x + 1, y + 1) + \mu v(x + 1, y + 1) + \nu}{(\mu + \nu)u(x, y) + (1 - \nu)v(x, y) + \lambda + 2}$$

taking $\lambda = 1$.

Write $k = \mu + \nu, \; k\mu = 1 - \nu, \; k\nu = u + 2,$

so that
$$\begin{vmatrix} k & -1 & -1 \\ - & k & 1 \\ -2 & -1 & k \end{vmatrix} = 0.$$

Thus
$$k^3 - 2k + 1 = 0,$$

i.e. $k_1 = 1, \; k_2 = (-1 + \sqrt{5})/2, \; k_3 = -(1 + \sqrt{5})/2.$

The equations for μ and ν then give—

$$\begin{cases} \mu + \nu = 1, \\ \mu + 2 = \nu. \end{cases} \quad \begin{cases} \mu + \nu = (-1 + \sqrt{5})/2, \\ \mu + 2 = (-1 + \sqrt{5})\nu/2. \end{cases} \quad \begin{cases} \mu + \nu = -(1 + \sqrt{5})/2, \\ \mu + 2 = -(1 + \sqrt{5})\nu/2. \end{cases}$$

These lead to

$$\begin{aligned} &\mu_1 = 1/2, && \nu_1 = 3/2, && k_1 = 1, \\ &\mu_2 = -1, && \nu_2 = (1 + \sqrt{5})/2, && k_2 = (-1 + \sqrt{5})/2, \\ &\mu_3 = -1, && \nu_3 = (1 - \sqrt{5})/2, && k_3 = -(1 + \sqrt{5})/2, \end{aligned}$$

giving finally

$$\frac{u(x, y) - \frac{1}{2}v(x, y) + 3/2}{u(x, y) - v(x, y) + (1 + \sqrt{5})/2} = \left(-\frac{1 + \sqrt{5}}{2}\right)^x \phi_1(x - y)$$

and

$$\frac{u(x, y) - \frac{1}{2}v(x, y) + 3/2}{u(x, y) - v(x, y) + \dfrac{1 - \sqrt{5}}{2}} = \left(-\frac{1 - \sqrt{5}}{2}\right)^x \phi_2(x - y),$$

from which $u(x, y)$ and $v(x, y)$ follow immediately in terms of the two arbitrary functions $\phi_1(x - y)$ and $\phi_2(x - y)$.

EXERCISES (XII)

1. $\dfrac{u(x + 1, y + 1)}{u(x, y) - v(x, y) + 1} = \dfrac{v(x + 1, y + 1)}{u(x, y) + v(x, y) - 2} = \dfrac{1}{u(x, y) - v(x, y)}.$

2. $\dfrac{u(x + 1, y + 1)}{2v(x, y) + 3} = \dfrac{v(x + 1, y + 1)}{u(x, y) + 1} = \dfrac{1}{u(x, y) - v(x, y)}.$

(*f*) *Functional Equations* in which like independent variables do not differ by an integral number of steps, may frequently be reduced by a transformation of the independent variables.

Example

$$z(ax + b, cy + d) = pz(Ax + B, Cy + D)^n.$$

Let

$$ax + b = u(t), \qquad cy + d = v(s),$$

$$Ax + B = u(t + 1), \qquad Cy + D = v(s + 1),$$

i.e.

$$au(t + 1) - Au(t) = aB - Ab,$$

$$cv(s + 1) - Cv(s) = cD - dC.$$

These two equations enable $u(t)$ and $v(s)$ to be found, so that x and y are expressed in terms of the new variables t and s. We notice that *any particular solution* for $u(t)$ and $v(s)$ will suffice. The original equation then takes the more orthodox form

$$z[u(t), v(s)] = pz[u(t + 1), v(s + 1)]^n$$

or

$$z_{t,\,s} = pz^n_{t+1,\,s+1}$$

and z is derived at once in terms of s and t with two arbitrary periodics, and therefore also in terms of x and y.

EXERCISES (XIII)

1. $z(x + 3, 2y + 1) = 2z(2x + 1, y + 3).$
2. $z(x + 1, y) = 3z(2x - 1, y^2).$

3. $z\left(\frac{4x+2}{x+3}, \frac{y-2}{y+4}\right) = 4z(x, y).$

The solution of certain types of functional equation involving several variables can be shown to be dependent on the solution of a system of partial difference equations. Thus suppose

$$z[\phi(u, v, w), \psi(u, v, w), \chi(u, v, w)] = az(u, v, w) + b,$$

where a and b are constants and ϕ, ψ, χ are given functions of u, v, and w the independent variables. To determine the form of $z(u, v, w)$ we write

$$u = u(r, s, t), \quad v = v(r, s, t), \quad w = w(r, s, t),$$

r, s, t being the new independent variables, and

$$u(r+1, s+1, t+1) = \phi[u(r, s, t), v(r, s, t), w(r, s, t)]$$
$$v(r+1, s+1, t+1) = \psi[u(r, s, t), v(r, s, t), w(r, s, t)]$$
$$w(r+1, s+1, t+1) = \chi[u(r, s, t), v(r, s, t), w(r, s, t)].$$

The latter three equations are partial difference equations determining the transformations of u, v, w in terms of r, s, and t. For this purpose the general solution is not required, any special set of expressions for u, v, w, satisfying the equations will suffice.

Now the original equation in z may be written

$$z_{r+1,\, s+1,\, t+1} = az_{r,\, s,\, t} + b$$

which is of the form $\quad (\mathrm{EFG} - a)z_{r,\, s,\, t} = b$

where E, F, G operate on r, s, t respectively in the usual way. The solution of this equation provides z in terms of r, s, and t with the appropriate number of arbitrary functions. The variables r, s, and t when eliminated from the expressions in terms of those for u, v, w, and z, finally provide the expression for z in terms of u, v, and w as required.

For example, given the relation

$$\lambda z(u, v) = z(au + bv + c, Au + Bv + C)$$

it is required to find the explicit expression for $z(u, v)$. Write

$$u = u(r, s) \qquad v = v(r, s)$$
$$u(r+1, s+1) = au(r, s) + bv(r, s) + c$$
$$v(r+1, s+1) = Au(r, s) + Bv(r, s) + C.$$

The three equations for u, v, and z in terms of r and s becomes—

$$z[u(r+1, s+1), v(r+1, s+1)] = \lambda z[u(r, s), v(r, s)]$$
$$(\mathrm{EF} - a)u(r, s) - bv(r, s) = c$$
$$Au(r, s) - (\mathrm{EF} - B)v(r, s) = C.$$

The first of these gives

$$z(u, v) = \lambda^r G(r - s),$$

where G is arbitrary and the remaining two equations are simple linear partial difference equations in the combined operator EF and so easily solved by the methods already outlined. Any special solution that involves r and s will suffice.

In illustration consider the following.

$$z(7u + 9v, 4u + 7v) = 13z(u, v)$$

Let
$$u(s+1, t+1) = 7u(s, t) + 9v(s, t)$$
$$v(s+1, t+1) = 4u(s, t) + 7v(s, t).$$

Thus
$$(\mathrm{EF} - 7)u(s, t) - 9v(s, t) = 0$$
$$-4u(s, t) + (\mathrm{EF} - 7)v(s, t) = 0,$$

so that
$$[(\mathrm{EF} - 7)^2 - 36]v(s, t) = 0$$

or
$$(\mathrm{EF} - 13)(\mathrm{EF} - 1)v(s, t) = 0.$$

In general therefore
$$v(s, t) = f(s - t) + 13^t g(s - t).$$

As we require only a particular transformation we take

$$v(s, t) = 13^{s-t}[1 + 13^t] = 13^{s-t} + 13^s$$

so that
$$u(s, t) = \tfrac{3}{2}[(-13)^{s-t} + 13^s].$$

Since the original equation can be written

$$z[u(s+1, t+1), v(s+1, t+1)] = 13z[u(s, t), v(s, t)]$$

it follows that $z(u, v) = 13^s F(s - t)$ where F is an arbitrary function. Hence from the expressions for u and v we have

$$(3v + 2u)/3 = 2 \,.\, 13^s$$

and
$$(3v - 2u)/3 = 2 \,.\, 13^{s-t},$$

i.e.
$$13^s = \frac{3v + 2u}{6},$$

and $s - t$ is a simple function of $3v - 2u$. Thus

$$z(u, v) = (3v + 2u)G(3v - 2u)$$

where G is arbitrary.

Check

$$\begin{aligned} z(7u + 9v, 4u + 7v) &= [3(4u + 7v) + 2(7u + 9v)]G[3(4u + 7v) \\ &\qquad - 2(7u + 9v)] \\ &= (26u + 39v)G(-2u + 3v) \\ &= 13(2u + 3v)G(3v - 2u) \\ &= 13z(u, v). \end{aligned}$$

EXERCISES (XIV)

Determine $z(x, y)$ in each of the following—

1. $z(x - y + 1, 2x + 4y) = 6z(x, y)$,
2. $z(x - y + 1, 2x + 4y) = [z(x, y)]^6$.

Determination of the Arbitrary Function in the Solution

At the present stage of this subject's development it is not possible to lay down any general method of resolving this problem since the variables in the arbitrary function may appear in a variety of forms. Accordingly this section will be devoted to a study of a range of special cases.

(*a*) $$u(x + 1, y) = au(x, y + 1) + bu(x, y)$$

where x is a positive integer, and where $u(0, y) = f(y)$, a given function of y.

The general solution to this equation has already been found, viz.

$$u(x, y) = b^x\left[A(y) + \frac{ax}{b}A(y + 1) + \frac{a^2x(x - 1)}{b^2 2!}A(y + 2) + \ldots\right].$$

Clearly on inserting $x = 0$ in this expression we find $A(y) = f(y)$ and the required solution is

$$u(x, y) = b^x\left[f(y) + \frac{ax}{b}f(y + 1) + \frac{a^2x(x - 1)}{b^2 2!}f(y + 2) + \ldots\right].$$

Since x is an integer the series terminates.

EXERCISE (XV)

The equation $0 = u(x + 1, y) - 2u(x, y + 1) + 2u(x, y)$ where x is a positive integer is such that u is subject to the condition that when $x = 0$, $u = c^{-y}$. Show that $u(x, y) = 2^x(1 + c)^x/c^{x+y}$.

(*b*) $$\alpha u(x, y) = u(x + 1, y) - u(x, y + 1)$$

where x is an integer, and where

$$u = pa^y + qb^y + rc^y + \ldots$$

along $x = 0$. Using Lagrange's method of solution we obtain—

$$u(x, y) = \sum_n A_n(\mu_n + \alpha)^x \mu_n^{\,y},$$

and accordingly

$$pa^y + qb^y + rc^y + \ldots = \sum_n A_n \mu_n^{\,y}$$

Thus

$$p = A_1,\ q = A_2,\ r = A_3 \ldots \mu_1 + \alpha = a,\ \mu_2 + \alpha = b \ldots$$

so that $u(x, y) = p(\alpha + a)^x a^y + q(\alpha + b)^x b^y + \ldots.$

This is, of course, a special case of *Example* 1.

EXERCISE (XVI)

Solve the equation $0 = u(m + 1, n) - au(m, n + 1) - bu(m, n)$ subject to the condition that when $m = n$, $u = p^n$ where m and n are positive integers.

(*c*) $$0 = u(m + 1, n) - u(m, n + 2).$$

This is of the first order in m, but of the second in n. Hence the boundary conditions may either provide the value of u for all values of n at a given value of m; or the values of the function u for all values of m at two distinct values of n.

(*i*) $$0 = u(m + 1, n) - u(m, n + 2)$$

$$u = f(n) \quad \text{when } m = 0.$$

Write the equation $\quad 0 = (\mathrm{E} - \mathrm{F}^2)u(m, n)$,

i.e. $$u(m, n) = \mathrm{F}^{2m}A(n) = A(n + 2m).$$

Accordingly inserting $m = 0$

$$u(0, n) = f(n) = A(n)$$

and $u = f(n + 2m)$ is the required solution.

(*ii*) $$u(m+1, n) - u(m, n+2) = 0.$$

Let
$$u = f(m) \text{ when } n = 0$$
$$u = g(m) \text{ when } n = 1.$$

Write the equation

$$(\mathrm{F} - \mathrm{E}^{\frac{1}{2}})(\mathrm{F} + \mathrm{E}^{\frac{1}{2}})u(m, n) = 0$$
$$u(m, n) = \mathrm{E}^{n/2}A(m) + (-1)^n\mathrm{E}^{n/2}B(m)$$
$$= A(m + n/2) + (-1)^n B(m + n/2).$$

Inserting the boundary conditions—

$$f(m) = A(m) + B(m)$$
$$g(m) = A(m + \tfrac{1}{2}) - B(m + \tfrac{1}{2})$$

or
$$g(m - \tfrac{1}{2}) = A(m) - B(m)$$

Hence
$$2A(m) = f(m) + g(m - \tfrac{1}{2})$$
$$2B(m) = f(m) - g(m - \tfrac{1}{2})$$

and accordingly

$$2u(m, n) = [1 + (-1)^n]f\left(m + \frac{n}{2}\right) + [1 - (-1)^n]g\left(m + \frac{n}{2} - \frac{1}{2}\right)$$

(*d*) To determine a solution of the equation

$$0 = z(x + 1, y + 1) - az(x, y),$$

which contains the curve

$$x = P(t),\ y = Q(t),\ z = R(t).$$

The general solution of the equation is

$$z = a^x\phi(y - x)$$

where ϕ is arbitrary. Hence in order that the given curve may lie on this surface

$$R(t) = a^{P(t)}\phi[Q(t) - P(t)].$$

Write $s = Q(t) - P(t)$, and express $R(t)a^{-P(t)}$ in terms of s, say $\psi(s)$. Hence $\phi(s) = \psi(s)$ and the solution required is

$$z = a^x\psi(y - x).$$

For example if the curve be

$$x = \cos t = P(t), \quad y = \sin t = Q(t), \quad z = \sin 2t = R(t),$$

then
$$\sin 2t = a^{\cos t}\,\phi[\sin t - \cos t]$$

Writing $s = \sin t - \cos t$, it follows that

$$\sin 2t = 1 - s^2 \quad \text{and} \quad \cos t = \tfrac{1}{2}[-s + \sqrt{(2 - s^2)}];$$

and inserting in the general solution

$$\phi(s) = [1 - s^2]a^{\frac{1}{2}(s - \sqrt{(2-s^2)})}.$$

Accordingly $$z = [1 - (x - y)^2]a^{\frac{1}{2}[3x - y - \sqrt{(2-[x-y]^2)}]}.$$

EXERCISES (XVII)

1. Find the equation to the surface $z = \phi(x, y)$ which contains the ellipse $z = x - y$, $x^2 + y^2 = 1$, and satisfies the difference relation

$$4z(x, y) = z(x + 1, y + 1).$$

Also determine the surface containing the straight line

$$x = \tfrac{1}{2}y = \frac{z - 1}{0}$$

2. Solve $z(x + 2, y) - z(x + 1, y + 1) = 2[z(x + 1, y) - z(x, y + 1)]$ and find the surface satisfying this relation and containing the straight line $x = -y = -z$, and the parabola $z = x^2$, $y = 0$.

3. Find the surface $z = z(x, y)$ which satisfies the relation

$$z(x + 2, y) + abz(x, y + 1) = az(x + 1, y + 1) + bz(x + 1, y)$$

and contains the two parabolas $z = x^2$ in the $x - z$ plane, and $z = y^2$ in the $y - z$ plane.

Special Forms of Solution

When difference or functional equations are derived from a physical problem it is frequently the case that some information is available about the general structure of the solution although the latter is not known in detail. Here we shall merely illustrate the method of passing from the difference equation to the special form of solution.

$$(a)\quad z^2(x+1, y+1) + f(x+y)g(x-y)z(x, y+1)z(x+1, y) + h(x+y)z^2(x, y) = 0.$$

We seek a solution of the form

$$z = F(r)G(s)$$

where $$r = x + y,\ s = x - y.$$

Here
$$z(x + 1, y + 1) = F(r + 2)G(s)$$
$$z(x, y + 1) = F(r + 1)G(s - 1)$$
$$z(x + 1, y) = F(r + 1)G(s + 1).$$

Hence the equation becomes

$$F^2(r+2)G^2(s) + f(r)g(s)F(r+1)G(s-1)F(r+1)G(s+1) + h(r)G^2(s)F^2(r) = 0.$$

Divide throughout by $F^2(r+1)f(r)G^2(s)$,

$$\frac{F^2(r+2)}{F^2(r+1)f(r)} + g(s)\frac{G(s-1)G(s+1)}{G^2(s)} + \frac{h(r)F^2(r+1)}{f(r)F^2(r1)} = 0.$$

Since the first and last terms are functions of r only and the middle term a function of s only, and r and s vary independently, we may write

$$g(s)G(s-1)G(s+1) = \lambda G^2(s)$$

$$F^2(r+2) + \lambda f(r)F^2(r+1) + h(r)F^2(r) = 0.$$

The former is a linear equation in $[G(s+1)]/G(s)$, viz.

$$\frac{G(s+1)}{G(s)} = \frac{\lambda}{g(s)}\frac{G(s)}{G(s-1)},$$

and is immediately soluble in finite terms for $G(s)$. The latter is a linear equation in $F^2(r)$ and can be dealt with by the methods of the earlier chapters.

(*b*) To determine the solutions of

$$u(x+1, y) - au(x+1, y+1) + bu(x, y) = 0$$

which are of the form

$$u(x, y) = F(x)G(y).$$

Here $\quad F(x+1)G(y) - aF(x+1)G(y+1) + bF(x)G(y) = 0,$

and therefore

$$1 - aG(y+1)/G(y) + bF(x)/F(x+1) = 0.$$

Write $\quad G(y+1) = \lambda G(y), \quad F(x+1) = \mu F(x)$

where $\quad 1 - a\lambda + b/\mu = 0.$

Hence $\quad G(y) = A\lambda^y, \quad F(x) = B\mu^x$

where A and B are unit periodics in y and x respectively, and the required solution is

$$u(x, y) = AB\mu^x\left(\frac{1 + b/\mu}{a}\right)^y = AB\mu^{x-y}\left(\frac{\mu + b}{a}\right)^y$$

where μ is arbitrary.

This is of course the type of solution that would have been obtained directly by Lagrange's method.

(c) To determine the ways of passing in the x, y plane from (0, 0) to (x, y) always moving parallel to the positive direction of the co-ordinate axes by one unit.

Let $u(x, y)$ be the number. Then clearly $u(x, y) = u(y, x)$.

Also

$$u(x + 1, y + 1) = u(x + 1, y) + u(x, y + 1),$$

since $(x + 1, y + 1)$ can be reached only from $(x + 1, y)$ and $(x, y + 1)$.

The form suggests that solutions may exist of the type

$$u(x, y) = P(x + y)Q(x)Q(y).$$

This is symmetrical in x and y. If any of these functions are absent they will reduce to a constant. Thus

$$P(x + y + 2)Q(x + 1)Q(y + 1) = P(x + y + 1)Q(x + 1)Q(y) + P(x + y + 1)Q(x)Q(y + 1).$$

Hence
$$\frac{P(x + y + 2)}{P(x + y + 1)} = \frac{Q(y)}{Q(y + 1)} + \frac{Q(x)}{Q(x + 1)}.$$

Since the left-hand side is a function of $x + y$ it follows that

$$\frac{Q(x)}{Q(x + 1)} = x + \lambda, \quad \frac{Q(y)}{Q(y + 1)} = y + \lambda,$$

therefore
$$Q(x) = A\,\frac{1}{(x + \lambda - 1)(x + \lambda - 2)\ldots\lambda},$$

and
$$Q(y) = A\,\frac{1}{(y + \lambda - 1)(y + \lambda - 2)\ldots\lambda}.$$

Also
$$P(x + y + 2) = (x + y + 2\lambda)P(x + y + 1),$$

or
$$P(t) = (t + 2\lambda - 2)P(t - 1).$$

Hence

$$P(t) = B(t + 2\lambda - 2)\,.\,(t + 2\lambda - 3)\,.\,(t + 2\lambda - 4)\ldots 2\lambda$$

or
$$u(x, y) = C\,\frac{(x + y + 2\lambda - 2)\ldots 2\lambda}{(x + \lambda - 1)(x + \lambda - 2)\ldots\lambda(y + \lambda - 1)\ldots\lambda}$$

Now when $x = 0$, $u(0, y) = 1$,

$$1 = C\frac{(y + 2\lambda - 2) \dots (2\lambda)}{(y + \lambda - 1) \dots \lambda}.$$

which requires $\lambda = 1$ and $C = 1$. Hence

$$u(x, y) = \frac{(x + y)!}{x!y!}$$

a result that could have been obtained directly. The number of ways of passing in this way from one corner of a square of side N to the opposite corner is $(2N)!/(N!)^2$.

(*d*) $f(m)u(m + 1, n) - g(n)u(m, n + 1) = [f(m) - g(n)]u(m, n)$

Assume $$u(m, n) = M(m) \,.\, N(n),$$

then

$$f(m)M(m + 1)N(n) - g(n)M(m)N(m + 1) = [f(m) - g(n)]M(m)N(n),$$

i.e. $$f(m)\frac{M(m + 1)}{M(m)} - g(n)\frac{N(n + 1)}{N(n)} = f(m) - g(n),$$

or $$f(m)\left[\frac{M(m + 1)}{M(m)} - 1\right] = \left[\frac{N(n + 1)}{N(n)} - 1\right]g(n) = A,$$

where A is a constant. Thus $M(m)$ and $N(u)$ are derived from equations of similar form,

$$M(m + 1) = [1 + A/f(m)]M(m).$$

Hence $$M(m) = \prod_{m=1}^{m-1}[f(m) + A]/f(m)$$

so that

$$u(m, n) = B\{\prod_{m=1}^{m-1}[f(m) + A]/f(m)\}\{\prod_{n=1}^{n-1}[g(n) + A]/g(n)\}$$

and since the original equation is linear we may write

$$u(m, n) = \sum_r B_r \prod_{m=1}^{m-1}\frac{f(m) + A_r}{f(m)} \,.\, \prod_{n=1}^{n-1}\frac{g(n) + A_r}{g(n)}.$$

(*e*) $(m^2 - n^2)(m + n)z(m, n)$

$$= (m - n)(m - n + 1)z(m + 1, n) + z(m, n + 1)$$

subject to the conditions

$$z = 0, \; m = n; \; z = \sin \frac{m\pi}{2} \Big/ \Gamma(2m + 1), \; m = -n.$$

The boundary conditions and the structure of the equations suggest writing

$$u = m + n, \quad v = m - n,$$

and seeking solutions of the form

$$z = U(u)V(v).$$

The equation then takes the form

$$u^2 v U(u)V(v) = v(v + 1)U(u + 1)V(v + 1) + U(u + 1)V(v - 1),$$

and this leads to

$$\frac{u^2 U(u)}{U(u + 1)} = \frac{v(v + 1)V(v + 1) + V(v - 1)}{vV(v)} = 2a, \text{ say.}$$

Hence $\quad U(u + 1) = u^2 U(u)/2a \quad$ or $\quad U(u) = [\Gamma(u)]^2/(2a)^u,$

and $\qquad v(v + 1)V(v + 1) - 2avV(v) + V(v - 1) = 0.$

Raising v by unity and multiplying throughout by $\Gamma(v + 1)$ gives

$$\Gamma(v + 3)V(v + 2) - 2a\Gamma(v + 2)V(v + 1) + \Gamma(v + 1)V(v) = 0.$$

Thus $\Gamma(v + 1)V(v) = A[a + i\sqrt{1 - a^2}]^v + B[a - i\sqrt{1 - a^2}]^v$

$$= C \sin (v\theta + \alpha)$$

where C and α are arbitrary and $a = \cos \theta$.

Therefore $\qquad z = \Sigma C \dfrac{\Gamma(m + n)^2}{(2a)^{m+n}} \cdot \dfrac{\sin [(m - n)\theta + \alpha]}{\Gamma(m - n + 1)}$

summed over all possible values of a, α, and C, i.e.

$$z = \frac{\Gamma(m + n)^2}{2^{m+n}\Gamma(m - n + 1)} \sum \frac{C \sin [(m - n)\theta + \alpha]}{a^{m+n}}.$$

When $m = n$, $z = 0$, so that $\alpha = 0$.

When $\qquad m = -n, \qquad z = \left(\sin \dfrac{m\pi}{2}\right) \Big/ \Gamma(2m + 1),$

so that $\sin \frac{m\pi}{2} = \Sigma C \sin 2m\theta$ for all m.

Hence $\theta = \pi/4,\ C = 1$, i.e. $a = 1/\sqrt{2}$,

and
$$z(m, n) = \frac{\Gamma(m+n)^2 \sin (m-n)\pi/4}{2^{(m+n)/2}\Gamma(m-n+1)}.$$

EXERCISE (XVIII)

Show that the solution in the form $z(m, n) = \Sigma f(m)g(n)$ for the equation

$$2(m+1)nz(m, n+1) = n(n+1)z(m+1, n+2) + z(m+1, n)$$

is
$$z = \frac{m!}{(n-1)!} \sum_{A,\alpha,\theta} A \cos (n\theta + \alpha)/\cos^m \theta$$

where A, α, and θ are arbitrary constants.

Answers to Exercises

Chapter 1

1 (I)

1. $\Delta^2 y_n = 2$, $\Delta^3 y_n = 6$, $\Delta^4 y_n = 24$
2. $(n^2 + n)$, $n^3 + 2n^2 + n$, $n^4 + n + 1$
3. (i) $\Delta y_n = 2 \cos 3(n + 1/2) \sin (3/2)$
   (ii) $a^n(a - 1)[(a - 1)n + 2a]$, $6(n + 1)$, 0
6. $3n(3 - n)$, $-(2n + 1)$

1 (III)

1. $2n^2 - n + 1$
2. $6y_n = (3 - 2b)n^3 + 6bn^2 + (3 - 4b)n + 6$
   (where b is an arbitrary constant)

1 (VIII)

1. $x + x(x - 1) + 3x(x - 1)(x - 2)$
2. $1 + x + 4x(x - 1)$
3. $1 + 2x(x - 1)(x - 2)(x - 3)$

1 (XI)

1. (i) $y_n = \alpha n + A$
   (ii) $y_n = n^2 + n + A$
   (iii) $y_n = \alpha(1 - \alpha^{n-1})/(1 - \alpha) + A$
   (iv) $y_n = n(n - 1)(n - 2)/3 + A$
   (v) $y_n = A - [\sin (n - 1/2)]/2 \sin (1/2)$.

1 (XII)

1. $n(n - 1)(n - 2)(n + 1)/4$
2. $\sum_1^n na^n = na(1 - a^n)/(1 - a) - a\{n(1 - a) - (1 - a^n)\}/(1 - a)^2$

1 (XIII)

1. $a^n/(2 - 3a)$
2. $n(n - 1)/2$
3. $12(n^2 + 10n + 46)$

1 (XIV)

5. $n^2 + n - 1$, $-\frac{1}{2}\left(n^3 + \frac{3n^2}{2} + 6n + \frac{39}{4}\right)$
7. $(x - r)^2$

1 (XV)

1. $$\sum_{r=1}^{n} \frac{r(n + 1)n(n - 1) \ldots (n - r + 1)\Delta^{r-1}y_0}{(r + 1)!}$$
2. $\sum_1^n n^4 = n + 15n(n - 1)/2 + 25n(n - 1)(n - 2)/3$
   $+ 5n(n - 1)(n - 2)(n - 3)/2 + n(n - 1)(n - 2)(n\ 3)(- n - 4)/5$
3. $\sum_1^n n^2(n + 1) = 2n + 5n(n - 1) + 7n(n - 1)(n - 2)/3$
   $+ n(n - 1)(n - 2)(n - 3)/4$

1 (XVI)

1. (i) $(1+x)/(1-x)^3$, (ii) $2/1(-x)^3$

1 (XVIII)

1. $\cos x - \dfrac{3x}{2}\sin x$

2. $e^x(1+2x+x^2)$

3. $3\cosh x + x\sinh x$

1 (XIX)

3. $v(t) = a^t(1 - a\lambda)$

1 (XX)

2. $u^2_{n+r} - ru^2_{n+r-1} + \ldots + (-1)^r u^2_n$

3. $u_n\Delta^2 v_n + v_n\Delta^2 u_n + \Delta^2 u_n\Delta^2 v_n + 2(\Delta u_n\Delta v_n + \Delta^2 u_n\Delta v_n + \Delta u_n\Delta^2 v_n)$

Miscellaneous

4. $a_0 + (a_1 + a_2 + a_3)n + (a_2 + 3a_3)n(n-1) + a_3 n(n-1)(n-2)$;

$$\frac{a_0}{2}n(n-1) + \left(\frac{a_1 + a_2 + a_3}{3!}\right)n(n-1)(n-2) + \frac{2(a_2 + 3a_3)}{4!}n(n-1)(n-2)(n-3) + \frac{6a_3}{5!}n(n-1)(n-2)(n-3)(n-4) + An + B$$

5. $e^x(1 + 2x + 2x^2 + x^3)$

Chapter 2

2 (I)

1. 0·23409

2. 0·86337

2 (III)

1. 1·03667

2 (VI)

2. − 23,172

3. $y = 2^x + 3^x - 1$

Chapter 3

3 (IV)

1. $y_n^2 = A\,.\,2^n + B\,.\,3^n$

2. $y_n^3 = A/n$

3. $y_n = A/n^2$

4. $y_n = [(n-1)!]^2$

5. (i) $\log y_n = A\,.\,\lambda^n$

(ii) $y_n = A\,.\,2^n - \log\lambda$

Chapter 4

4 (I)

1. $D_n = \dfrac{1}{\sqrt{(\alpha^2 - 4)}}\left\{\left[\dfrac{\alpha + \sqrt{(\alpha^2-4)}}{2}\right]^{n+1} - \left[\dfrac{\alpha - \sqrt{(\alpha^2-4)}}{2}\right]^{n+1}\right\}$

4 (II)

1. (i) $A(x)2^x + B(x)(-1)^x$

(ii) $A(x)4^x + B(x)$

(iii) $[A(x) + B(x)(-1)^x]i^x$

(iv) $5^{x/2}[A(x)\cos(x\cot^{-1}2) + B(x)\sin(x\cot^{-1}2)]$
(v) $3^x[xA(x) + B(x)]$
(vi) $2^x[xA(x) + B(x)]$ where $A(x)$ and $B(x)$ are unit periodics

2. $y_n = 2^{2n-1} + (-2)^{n-1}$
3. $y_n = AB^n$
5. $y(x) = (Ax + B)/4^x$
6. (i) $(-1)^n[1/3 \cdot 4^{n-1} - 7/3]$
(ii) $(3n + 2)(-2)^{n-1}$

4 (III)

1. $y_n = A \cdot B^n$
2. $y_n = (An^2 + Bn + C)^{1/2}$
3. $1/y_n = A \cdot 3^n + B(-7)^n + e^n/(e + 7)(e - 3)$
4. $y_n = B - \sum_{1}^{n-1} 1/(n + A)$
5. $y_n = c + \sum_{1}^{n-1} 6/(n^3 - 3n^2 + An + B)$
6. $y_n = Ae^{Bn^3+Cn^2+Dn}$

4 (IV)

2. (i) $y_n = 1/3^n$
(ii) $y_n = 17 \cdot 3^n/36 - (2n + 1)/4$
(iii) $y_n = (4n - 7)2^{n-2}$
(iv) $y_n = n(n - 1)(2n - 1)2^{n-2}/3$
(v) $y(x) = 3^x + A(x)2^x + B(x)(-2)^x$
(vi) $y(x) = 2^x[x/8 + A(x) + B(x)(-1)^x]$
(vii) $y(x) = 2^{x-3}[x(x - 1) + xA(x) + B(x)]$

Chapter 5

5 (I)

7. 0·50634
11. 7363·11
17. $(2 - C^{2x})/(C^{2x} - 1)$

5 (III)

1. $y(x)/[q \cdot 5^x + B \cdot 2^x - A \cdot 3^x] = z(x)/[3 \cdot 5^x - A3^x/q]$
$= 1/[A \cdot 3^x - B \cdot 2^x - (q - 6)5^x]$
2. $y(x)/[A \cdot 2^x + rB \cdot 3^x + r \cdot 7^x] = z(x)/[-4A \cdot 2^x + 4rB \cdot 3^x + (20 - 4r) \cdot 7^x]$
$= 1/[5 \cdot 7^x + B \cdot 3^x]$

Chapter 6

6 (I)

1. $y_n = c[(n - 1)!]^2$
2. $y_n = c/n$
3. $y_n = Ca^{n^2}$
4. $y_n = c/a^{n^2}$
5. $y_n = c/n! + 1/n$
6. $y_n = n + ce^{n!}$
7. $y_n = n(n - 1) + cn^2$
8. $y_n = (n^2 - 1)/3 + c/n$
9. $y_n = e^{n(n-1)}[C + n(n - 1)(2n - 1)/2]$

6 (II)

1. $y(x) = C(x) + \dfrac{e^x}{e - 1}\left(x - \dfrac{e}{e - 1}\right)$

6 (III)

1. $y(x) = c(x) + x(x - 1)(2x - 1)/6$
2. $y(x) = x[C(x) + x(x - 1)(2x - 1)/6]$

3. $y(x) = C(x) - 1/x$
4. $y(x) = (-1)^x[C(x) - 1/x]$

6 (V)

3. $y(x) = A(x)\,2^x + xB(x) + 1$
4. (i) $y_n = (-1)^n A + B(-1)^n \sum_1^{n-1} (n-1)!^2$

(ii) $y_n = Aa^n + Ba^n \sum_1^{n-1} (-1)^n \left(1 + \frac{1}{a}\right)^n \Gamma\left(n + \frac{a}{a+1}\right) \Big/ \Gamma(n)$

(iii) $y_n = A + B \sum_1^{n-1} 1/(n-1)! + \sum_1^{n-1} \left(\sum_1^{n-1} n!\right) \Big/ (n-1)!$

(iv) $y_n = (-3)^n[A + B \sum_3^{n-1} 3^{-n}(n-3)!]$

(v) $y_n = A(-1)^n + B(-1)^n \sum_1^{n-1} (n-1)!$

(vi) $y_n/n(n-1)(n+1) = A + B \sum^{n-1} 3^n/(n+2)(n+1)n(n-1)$

7. $y_n/n = A + B \sum_1^{n-1} (-1)^n \Gamma(n+i)\Gamma(n-i)(n^2+1)/(n+2)!$

6 (VI)

1. (i) $y(x)/\Gamma(x) = A(-1)^x \sum_{n=1}^{N} (-1)^n/(x+n) + B$

(ii) $y(x)/\Gamma(x) = A \sum_{n=1}^{N} 1/(x-n) + B$

2. $p(x) = q(x) + A$

$$y(x) = B\,\frac{\Gamma(x-a_1)\Gamma(x-a_2)\ldots}{\Gamma(x-b_1)\Gamma(x-b_2)\ldots} + C\,\frac{\Gamma(x-A_1)\Gamma(x-A_2)\ldots}{\Gamma(x-B_1)\Gamma(x-B_2)\ldots}$$

where $p(x) = (x-a_1)(x-a_2)\ldots/(x-b_1)(x-b_2)\ldots$

$q(x) = (x-A_1)(x-A_2)\ldots/(x-B_1)(x-B_2)\ldots$

3. (i) $y(x)/P(x) = A + B\sum_0^\infty 1/a(x+n)$ where $P(x+1) = a(x)P(x)$

(ii) $y(x)/\Gamma(x) = A + B \sum_1^{N-1} 1/\Gamma([x]+n)\Gamma([x]+n+1)$ where N is the integer in x

(iii) $y(x) = (-1)^x \lambda^{x(x-3)/2}[Ax + B]$

(iv) $y(x)/\Gamma(x) = A(-1)^x + B(-1)^x \sum_0^{N-1} 1/([x]+n)$

6 (VII)

1. (i) $y(x) = \Gamma(x)\Gamma(x+1)\left[A + B\sum_1^N 1/\Gamma([x]+N)\right]$

(ii) $y(x) = [A + B(-1)^x]/\Gamma(x+1)$

(iii) $y(x) = a^{-x}(-1)^x\left[A + B\sum_x^\infty a^{x(x-1)/2}\right]$

(iv) $y(x) = A(-1)^{2x}[x - (x+1)/x^2 + (x+2)/x^2(x+1)^2 - (x+3)/x^2(x+1)^2(x+2)^2 + \ldots] + B(-1)^x\Gamma(x)^2$

(v) $y(x) = [A + B(-1)^x e^{-x}]e^{x(x-1)/2}$

(vi)

$$y(x) = \left[A(-1)^x 2^x + B\sum_n^\infty 2^{-n}(-1)^n\Gamma(x+n)/\Gamma(x+n+\tfrac{1}{2})\right]\Gamma(x-\tfrac{1}{2})/\Gamma(x)$$

2. (iii) $y_n = 3^{-n}\Gamma(n-1)\left[A + B\sum_{n}^{\infty} 1/n\Gamma(n)\right]$

(iv) $x + 2$

(v) $\lambda = 3,\ a = 1/9,\ b = 4/9$

(vi) $y_n = y_0 + An + Bn^2$

4. $y(x) = [A\alpha^x + B\beta^x]\prod^{x-2} p(x)$ where $\alpha^2 + a\alpha + 1 = \beta^2 + a\beta + 1 = 0$,
$y(x) = [A\alpha^x + B\beta^x]\,\Gamma(x-1)$ where $\alpha^2 + \alpha + 1 = \beta^2 + \beta + 1 = 0$

5. $y(x) = \left[A + B\sum_{n=0}^{N-1} \Gamma\{[x] + n\}\right]\Gamma(x)(-1)^x$ where $x = [x] + N$

6 (IX)

1. $y(x) = 2^x[A + Bx^2]$

2. $y_n = An + B/n$

6 (XV)

1. $$y(x) = A\left[1 - \frac{a}{(x+1)} + \frac{a^2}{2!(x+1)(x+2)} - \frac{a^3}{3!(x+1)(x+2)(x+3)}\right]$$

2. $$J_n(x) = \frac{x^n}{2^n n!}\left[1 - \frac{x^2}{2^2(n+1)} + \frac{x^4}{2!\,2^4(n+1)(n+2)} + \cdots\right]$$

3. $$y(x) = A\left[1 + \frac{\lambda x(x+1)}{2^2} + \frac{\lambda^2 x(x+1)(x+2)(x+3)}{2^2 . 4^2} + \cdots\right]$$

5. $$y(x) = Ax(x+1)\left[1 - \frac{x+2}{1 . 5} + \frac{(x+2)(x+3)}{1 . 2 . 5 . 6} - \frac{(x+2)(x+3)(x+4)}{1 . 2 . 3 . 5 . 6 . 7} + \cdots\right]$$

6. $$y(x) = A\left[x + \frac{x(x-1)}{(2^2-1)} + \frac{x(x-1)(x-2)}{(2^2-1)(3^2-1)} + \cdots\right]$$

7. $$y(x) = A\left[\frac{1}{x} - \frac{\lambda}{x(x+1) . 1!} + \frac{\lambda^2}{x(x+1)(x+2)2!} - \cdots\right]$$

6 (XVIII)

1. (i) Between 2 and 3, (ii) Between 2 and 3, (iii) Between 2 and 4

2. $0{\cdot}25 < \lambda < 0{\cdot}39$

Chapter 8

8 (II)

3. $(-1)^{m+n}[u(x, y) - \{mu\,(x+1, y) + nu\,(x, y+1)\}$
$+ \{m\,(m-1)u(x+2, y) + 2mnu\,(x+1, y+1)$
$+ n(n-1)u(x, y+2)\}/2! + \ldots]$

4. $n(x, y) + (m\Delta_x + n\Delta_y)u(x, y) + \frac{1}{2}!\{m(m-1)\Delta^2_x + 2mn\Delta_x\Delta_y$
$+ n(n-1)\Delta^2_y\}\,u(x, y) + \ldots$

8 (III)

1. $$u(x, y) = A(x) + yA(x+1) + \frac{y(y-1)A(x+2) + \ldots}{2!} + \frac{y[B(x) + yB(x+1) + y(y-1)B(x+2)]}{2!}$$

8 (V)

1. $B(y) + (-1)^x c(y) + \frac{x}{2}[B(y+1) + (-1)^x c(y+1)] + \ldots,$

or $(-1)^y[A\,(x) - yA(x+2) + \dfrac{y(y-1)A(x+4)}{2!} - \ldots$

8 (VI)

2. $u(x, y) = \Sigma\alpha^x_r A_r(y - x)$ where $\phi(\alpha_r) = 0$, A_r = unit periodic; corresponding to $(\mathrm{EF} - \alpha)^n$ there is a term $\alpha^x \sum_{0}^{n-1} x^r A_r(y - x)$

8 (VIII)

1. $2^x e^{-2y} A(x + y) + \dfrac{e^{x-2y}(x + y - 1)}{1 - 2e^{-1}}$
2. $x(x - 1)(x + y - 2)/2 + A(x + y) + xB(x + y)$
3. $A(x + y) + 2^x B(x + y) + 2^x x(x + 2y - 3)/4$
4. $4^x A(x - y) + 5^x B(x - y) + xy/12 + 7(x + y)/144 + 79\,864$
5. $x(x - y)^2 + A(x - y)$
6. $2^x(x - y)[A(x - y) + (x + y)(x + y - 1)(2x + 2y - 1)/12]$

8 (IX)

1. $e^{p^x} A(x + y)$

8 (X)

1. $z = Ax + By + AB$

(XII)

1. $u(x, y)/[6 \,.\, 2^x + 2A(x - y)]$
   $= v(x, y)/[-2 \,.\, 2^x + A(x - y) + (-1)^x B(x - y)]$
   $= 1/(4 \,.\, 2^x + A(x - y) + (-1)^x B(x - y)]$

8 (XIII)

1. $z(x, y) = \mathrm{F}\{(x - 5)(y - 5)\}/(x - 5)$
2. $z(x, y) = \mathrm{F}\{(\log y)/(x - 3)\} \,.\, 3^{-\log(x-3)/\log 2}$
3. $z(x, y) = \left(\dfrac{y + 1}{y + 2}\right)^{\log 4/\log 2/3} \mathrm{F}\left[\left(\dfrac{x + 1}{x - 2}\right)^{\log 2/3}\left(\dfrac{y + 1}{y + 2}\right)^{\log 2/5}\right]$

8 (XIV)

1. $z = (2x + 2y + 1)(2x + y + 2)\mathrm{F}\{(2x + y + 2)^{\log 3}/(2x + 2y + 1)^{\log 2}\}$
2. $\log z$ = same function as in question 1

8 (XVI)

1. $u = p^{(m+n)/2}[A\alpha^{m-n} + B\beta^{m-n}]$ where $A + B = 1$, and α and β are roots of $x^2\sqrt{p} = bx + a\sqrt{p}$

8 (XVII)

1. $\log z = \log(x - y) + \{y \pm \sqrt{[2 - (x - y)^2]}\} \log 4$
   $\log z = (2x - y) \log 4$
2. $z = (x - y)^2 + y(1 - 4y)2^{x+y}$
3. $z = b^x(1 - a^y)y^2 - Aa^y b^{x-y} + (x - y)^2 a^y$, where A is a constant

Index

A CATALOG OF SELECTED

DOVER BOOKS

IN SCIENCE AND MATHEMATICS

Astronomy

CHARIOTS FOR APOLLO: The NASA History of Manned Lunar Spacecraft to 1969, Courtney G. Brooks, James M. Grimwood, and Loyd S. Swenson, Jr. This illustrated history by a trio of experts is the definitive reference on the Apollo spacecraft and lunar modules. It traces the vehicles' design, development, and operation in space. More than 100 photographs and illustrations. 576pp. 6 3/4 x 9 1/4. 0-486-46756-2

EXPLORING THE MOON THROUGH BINOCULARS AND SMALL TELESCOPES, Ernest H. Cherrington, Jr. Informative, profusely illustrated guide to locating and identifying craters, rills, seas, mountains, other lunar features. Newly revised and updated with special section of new photos. Over 100 photos and diagrams. 240pp. 8 1/4 x 11. 0-486-24491-1

WHERE NO MAN HAS GONE BEFORE: A History of NASA's Apollo Lunar Expeditions, William David Compton. Introduction by Paul Dickson. This official NASA history traces behind-the-scenes conflicts and cooperation between scientists and engineers. The first half concerns preparations for the Moon landings, and the second half documents the flights that followed Apollo 11. 1989 edition. 432pp. 7 x 10. 0-486-47888-2

APOLLO EXPEDITIONS TO THE MOON: The NASA History, Edited by Edgar M. Cortright. Official NASA publication marks the 40th anniversary of the first lunar landing and features essays by project participants recalling engineering and administrative challenges. Accessible, jargon-free accounts, highlighted by numerous illustrations. 336pp. 8 3/8 x 10 7/8. 0-486-47175-6

ON MARS: Exploration of the Red Planet, 1958-1978--The NASA History, Edward Clinton Ezell and Linda Neuman Ezell. NASA's official history chronicles the start of our explorations of our planetary neighbor. It recounts cooperation among government, industry, and academia, and it features dozens of photos from Viking cameras. 560pp. 6 3/4 x 9 1/4. 0-486-46757-0

ARISTARCHUS OF SAMOS: The Ancient Copernicus, Sir Thomas Heath. Heath's history of astronomy ranges from Homer and Hesiod to Aristarchus and includes quotes from numerous thinkers, compilers, and scholasticists from Thales and Anaximander through Pythagoras, Plato, Aristotle, and Heraclides. 34 figures. 448pp. 5 3/8 x 8 1/2. 0-486-43886-4

AN INTRODUCTION TO CELESTIAL MECHANICS, Forest Ray Moulton. Classic text still unsurpassed in presentation of fundamental principles. Covers rectilinear motion, central forces, problems of two and three bodies, much more. Includes over 200 problems, some with answers. 437pp. 5 3/8 x 8 1/2. 0-486-64687-4

BEYOND THE ATMOSPHERE: Early Years of Space Science, Homer E. Newell. This exciting survey is the work of a top NASA administrator who chronicles technological advances, the relationship of space science to general science, and the space program's social, political, and economic contexts. 528pp. 6 3/4 x 9 1/4. 0-486-47464-X

STAR LORE: Myths, Legends, and Facts, William Tyler Olcott. Captivating retellings of the origins and histories of ancient star groups include Pegasus, Ursa Major, Pleiades, signs of the zodiac, and other constellations. "Classic." – *Sky & Telescope.* 58 illustrations. 544pp. 5 3/8 x 8 1/2. 0-486-43581-4

A COMPLETE MANUAL OF AMATEUR ASTRONOMY: Tools and Techniques for Astronomical Observations, P. Clay Sherrod with Thomas L. Koed. Concise, highly readable book discusses the selection, set-up, and maintenance of a telescope; amateur studies of the sun; lunar topography and occultations; and more. 124 figures. 26 halftones. 37 tables. 335pp. 6 1/2 x 9 1/4. 0-486-42820-6

Chemistry

MOLECULAR COLLISION THEORY, M. S. Child. This high-level monograph offers an analytical treatment of classical scattering by a central force, quantum scattering by a central force, elastic scattering phase shifts, and semi-classical elastic scattering. 1974 edition. 310pp. 5 3/8 x 8 1/2. 0-486-69437-2

HANDBOOK OF COMPUTATIONAL QUANTUM CHEMISTRY, David B. Cook. This comprehensive text provides upper-level undergraduates and graduate students with an accessible introduction to the implementation of quantum ideas in molecular modeling, exploring practical applications alongside theoretical explanations. 1998 edition. 832pp. 5 3/8 x 8 1/2. 0-486-44307-8

RADIOACTIVE SUBSTANCES, Marie Curie. The celebrated scientist's thesis, which directly preceded her 1903 Nobel Prize, discusses establishing atomic character of radioactivity; extraction from pitchblende of polonium and radium; isolation of pure radium chloride; more. 96pp. 5 3/8 x 8 1/2. 0-486-42550-9

CHEMICAL MAGIC, Leonard A. Ford. Classic guide provides intriguing entertainment while elucidating sound scientific principles, with more than 100 unusual stunts: cold fire, dust explosions, a nylon rope trick, a disappearing beaker, much more. 128pp. 5 3/8 x 8 1/2. 0-486-67628-5

ALCHEMY, E. J. Holmyard. Classic study by noted authority covers 2,000 years of alchemical history: religious, mystical overtones; apparatus; signs, symbols, and secret terms; advent of scientific method, much more. Illustrated. 320pp. 5 3/8 x 8 1/2. 0-486-26298-7

CHEMICAL KINETICS AND REACTION DYNAMICS, Paul L. Houston. This text teaches the principles underlying modern chemical kinetics in a clear, direct fashion, using several examples to enhance basic understanding. Solutions to selected problems. 2001 edition. 352pp. 8 3/8 x 11. 0-486-45334-0

PROBLEMS AND SOLUTIONS IN QUANTUM CHEMISTRY AND PHYSICS, Charles S. Johnson and Lee G. Pedersen. Unusually varied problems, with detailed solutions, cover of quantum mechanics, wave mechanics, angular momentum, molecular spectroscopy, scattering theory, more. 280 problems, plus 139 supplementary exercises. 430pp. 6 1/2 x 9 1/4. 0-486-65236-X

ELEMENTS OF CHEMISTRY, Antoine Lavoisier. Monumental classic by the founder of modern chemistry features first explicit statement of law of conservation of matter in chemical change, and more. Facsimile reprint of original (1790) Kerr translation. 539pp. 5 3/8 x 8 1/2. 0-486-64624-6

MAGNETISM AND TRANSITION METAL COMPLEXES, F. E. Mabbs and D. J. Machin. A detailed view of the calculation methods involved in the magnetic properties of transition metal complexes, this volume offers sufficient background for original work in the field. 1973 edition. 240pp. 5 3/8 x 8 1/2. 0-486-46284-6

GENERAL CHEMISTRY, Linus Pauling. Revised third edition of classic first-year text by Nobel laureate. Atomic and molecular structure, quantum mechanics, statistical mechanics, thermodynamics correlated with descriptive chemistry. Problems. 992pp. 5 3/8 x 8 1/2. 0-486-65622-5

ELECTROLYTE SOLUTIONS: Second Revised Edition, R. A. Robinson and R. H. Stokes. Classic text deals primarily with measurement, interpretation of conductance, chemical potential, and diffusion in electrolyte solutions. Detailed theoretical interpretations, plus extensive tables of thermodynamic and transport properties. 1970 edition. 590pp. 5 3/8 x 8 1/2. 0-486-42225-9

Engineering

FUNDAMENTALS OF ASTRODYNAMICS, Roger R. Bate, Donald D. Mueller, and Jerry E. White. Teaching text developed by U.S. Air Force Academy develops the basic two-body and n-body equations of motion; orbit determination; classical orbital elements, coordinate transformations; differential correction; more. 1971 edition. 455pp. 5 3/8 x 8 1/2. 0-486-60061-0

INTRODUCTION TO CONTINUUM MECHANICS FOR ENGINEERS: Revised Edition, Ray M. Bowen. This self-contained text introduces classical continuum models within a modern framework. Its numerous exercises illustrate the governing principles, linearizations, and other approximations that constitute classical continuum models. 2007 edition. 320pp. 6 1/8 x 9 1/4. 0-486-47460-7

ENGINEERING MECHANICS FOR STRUCTURES, Louis L. Bucciarelli. This text explores the mechanics of solids and statics as well as the strength of materials and elasticity theory. Its many design exercises encourage creative initiative and systems thinking. 2009 edition. 320pp. 6 1/8 x 9 1/4. 0-486-46855-0

FEEDBACK CONTROL THEORY, John C. Doyle, Bruce A. Francis and Allen R. Tannenbaum. This excellent introduction to feedback control system design offers a theoretical approach that captures the essential issues and can be applied to a wide range of practical problems. 1992 edition. 224pp. 6 1/2 x 9 1/4. 0-486-46933-6

THE FORCES OF MATTER, Michael Faraday. These lectures by a famous inventor offer an easy-to-understand introduction to the interactions of the universe's physical forces. Six essays explore gravitation, cohesion, chemical affinity, heat, magnetism, and electricity. 1993 edition. 96pp. 5 3/8 x 8 1/2. 0-486-47482-8

DYNAMICS, Lawrence E. Goodman and William H. Warner. Beginning engineering text introduces calculus of vectors, particle motion, dynamics of particle systems and plane rigid bodies, technical applications in plane motions, and more. Exercises and answers in every chapter. 619pp. 5 3/8 x 8 1/2. 0-486-42006-X

ADAPTIVE FILTERING PREDICTION AND CONTROL, Graham C. Goodwin and Kwai Sang Sin. This unified survey focuses on linear discrete-time systems and explores natural extensions to nonlinear systems. It emphasizes discrete-time systems, summarizing theoretical and practical aspects of a large class of adaptive algorithms. 1984 edition. 560pp. 6 1/2 x 9 1/4. 0-486-46932-8

INDUCTANCE CALCULATIONS, Frederick W. Grover. This authoritative reference enables the design of virtually every type of inductor. It features a single simple formula for each type of inductor, together with tables containing essential numerical factors. 1946 edition. 304pp. 5 3/8 x 8 1/2. 0-486-47440-2

THERMODYNAMICS: Foundations and Applications, Elias P. Gyftopoulos and Gian Paolo Beretta. Designed by two MIT professors, this authoritative text discusses basic concepts and applications in detail, emphasizing generality, definitions, and logical consistency. More than 300 solved problems cover realistic energy systems and processes. 800pp. 6 1/8 x 9 1/4. 0-486-43932-1

THE FINITE ELEMENT METHOD: Linear Static and Dynamic Finite Element Analysis, Thomas J. R. Hughes. Text for students without in-depth mathematical training, this text includes a comprehensive presentation and analysis of algorithms of time-dependent phenomena plus beam, plate, and shell theories. Solution guide available upon request. 672pp. 6 1/2 x 9 1/4. 0-486-41181-8

HELICOPTER THEORY, Wayne Johnson. Monumental engineering text covers vertical flight, forward flight, performance, mathematics of rotating systems, rotary wing dynamics and aerodynamics, aeroelasticity, stability and control, stall, noise, and more. 189 illustrations. 1980 edition. 1089pp. 5 5/8 x 8 1/4. 0-486-68230-7

MATHEMATICAL HANDBOOK FOR SCIENTISTS AND ENGINEERS: Definitions, Theorems, and Formulas for Reference and Review, Granino A. Korn and Theresa M. Korn. Convenient access to information from every area of mathematics: Fourier transforms, Z transforms, linear and nonlinear programming, calculus of variations, random-process theory, special functions, combinatorial analysis, game theory, much more. 1152pp. 5 3/8 x 8 1/2. 0-486-41147-8

A HEAT TRANSFER TEXTBOOK: Fourth Edition, John H. Lienhard V and John H. Lienhard IV. This introduction to heat and mass transfer for engineering students features worked examples and end-of-chapter exercises. Worked examples and end-of-chapter exercises appear throughout the book, along with well-drawn, illuminating figures. 768pp. 7 x 9 1/4. 0-486-47931-5

BASIC ELECTRICITY, U.S. Bureau of Naval Personnel. Originally a training course; best nontechnical coverage. Topics include batteries, circuits, conductors, AC and DC, inductance and capacitance, generators, motors, transformers, amplifiers, etc. Many questions with answers. 349 illustrations. 1969 edition. 448pp. 6 1/2 x 9 1/4. 0-486-20973-3

BASIC ELECTRONICS, U.S. Bureau of Naval Personnel. Clear, well-illustrated introduction to electronic equipment covers numerous essential topics: electron tubes, semiconductors, electronic power supplies, tuned circuits, amplifiers, receivers, ranging and navigation systems, computers, antennas, more. 560 illustrations. 567pp. 6 1/2 x 9 1/4. 0-486-21076-6

BASIC WING AND AIRFOIL THEORY, Alan Pope. This self-contained treatment by a pioneer in the study of wind effects covers flow functions, airfoil construction and pressure distribution, finite and monoplane wings, and many other subjects. 1951 edition. 320pp. 5 3/8 x 8 1/2. 0-486-47188-8

SYNTHETIC FUELS, Ronald F. Probstein and R. Edwin Hicks. This unified presentation examines the methods and processes for converting coal, oil, shale, tar sands, and various forms of biomass into liquid, gaseous, and clean solid fuels. 1982 edition. 512pp. 6 1/8 x 9 1/4. 0-486-44977-7

THEORY OF ELASTIC STABILITY, Stephen P. Timoshenko and James M. Gere. Written by world-renowned authorities on mechanics, this classic ranges from theoretical explanations of 2- and 3-D stress and strain to practical applications such as torsion, bending, and thermal stress. 1961 edition. 560pp. 5 3/8 x 8 1/2. 0-486-47207-8

PRINCIPLES OF DIGITAL COMMUNICATION AND CODING, Andrew J. Viterbi and Jim K. Omura. This classic by two digital communications experts is geared toward students of communications theory and to designers of channels, links, terminals, modems, or networks used to transmit and receive digital messages. 1979 edition. 576pp. 6 1/8 x 9 1/4. 0-486-46901-8

LINEAR SYSTEM THEORY: The State Space Approach, Lotfi A. Zadeh and Charles A. Desoer. Written by two pioneers in the field, this exploration of the state space approach focuses on problems of stability and control, plus connections between this approach and classical techniques. 1963 edition. 656pp. 6 1/8 x 9 1/4. 0-486-46663-9

Mathematics–Bestsellers

HANDBOOK OF MATHEMATICAL FUNCTIONS: with Formulas, Graphs, and Mathematical Tables, Edited by Milton Abramowitz and Irene A. Stegun. A classic resource for working with special functions, standard trig, and exponential logarithmic definitions and extensions, it features 29 sets of tables, some to as high as 20 places. 1046pp. 8 x 10 1/2. 0-486-61272-4

ABSTRACT AND CONCRETE CATEGORIES: The Joy of Cats, Jiri Adamek, Horst Herrlich, and George E. Strecker. This up-to-date introductory treatment employs category theory to explore the theory of structures. Its unique approach stresses concrete categories and presents a systematic view of factorization structures. Numerous examples. 1990 edition, updated 2004. 528pp. 6 1/8 x 9 1/4. 0-486-46934-4

MATHEMATICS: Its Content, Methods and Meaning, A. D. Aleksandrov, A. N. Kolmogorov, and M. A. Lavrent'ev. Major survey offers comprehensive, coherent discussions of analytic geometry, algebra, differential equations, calculus of variations, functions of a complex variable, prime numbers, linear and non-Euclidean geometry, topology, functional analysis, more. 1963 edition. 1120pp. 5 3/8 x 8 1/2. 0-486-40916-3

INTRODUCTION TO VECTORS AND TENSORS: Second Edition--Two Volumes Bound as One, Ray M. Bowen and C.-C. Wang. Convenient single-volume compilation of two texts offers both introduction and in-depth survey. Geared toward engineering and science students rather than mathematicians, it focuses on physics and engineering applications. 1976 edition. 560pp. 6 1/2 x 9 1/4. 0-486-46914-X

AN INTRODUCTION TO ORTHOGONAL POLYNOMIALS, Theodore S. Chihara. Concise introduction covers general elementary theory, including the representation theorem and distribution functions, continued fractions and chain sequences, the recurrence formula, special functions, and some specific systems. 1978 edition. 272pp. 5 3/8 x 8 1/2. 0-486-47929-3

ADVANCED MATHEMATICS FOR ENGINEERS AND SCIENTISTS, Paul DuChateau. This primary text and supplemental reference focuses on linear algebra, calculus, and ordinary differential equations. Additional topics include partial differential equations and approximation methods. Includes solved problems. 1992 edition. 400pp. 7 1/2 x 9 1/4. 0-486-47930-7

PARTIAL DIFFERENTIAL EQUATIONS FOR SCIENTISTS AND ENGINEERS, Stanley J. Farlow. Practical text shows how to formulate and solve partial differential equations. Coverage of diffusion-type problems, hyperbolic-type problems, elliptic-type problems, numerical and approximate methods. Solution guide available upon request. 1982 edition. 414pp. 6 1/8 x 9 1/4. 0-486-67620-X

VARIATIONAL PRINCIPLES AND FREE-BOUNDARY PROBLEMS, Avner Friedman. Advanced graduate-level text examines variational methods in partial differential equations and illustrates their applications to free-boundary problems. Features detailed statements of standard theory of elliptic and parabolic operators. 1982 edition. 720pp. 6 1/8 x 9 1/4. 0-486-47853-X

LINEAR ANALYSIS AND REPRESENTATION THEORY, Steven A. Gaal. Unified treatment covers topics from the theory of operators and operator algebras on Hilbert spaces; integration and representation theory for topological groups; and the theory of Lie algebras, Lie groups, and transform groups. 1973 edition. 704pp. 6 1/8 x 9 1/4. 0-486-47851-3

Mathematics–Logic and Problem Solving

PERPLEXING PUZZLES AND TANTALIZING TEASERS, Martin Gardner. Ninety-three riddles, mazes, illusions, tricky questions, word and picture puzzles, and other challenges offer hours of entertainment for youngsters. Filled with rib-tickling drawings. Solutions. 224pp. 5 3/8 x 8 1/2. 0-486-25637-5

MY BEST MATHEMATICAL AND LOGIC PUZZLES, Martin Gardner. The noted expert selects 70 of his favorite "short" puzzles. Includes The Returning Explorer, The Mutilated Chessboard, Scrambled Box Tops, and dozens more. Complete solutions included. 96pp. 5 3/8 x 8 1/2. 0-486-28152-3

THE LADY OR THE TIGER?: and Other Logic Puzzles, Raymond M. Smullyan. Created by a renowned puzzle master, these whimsically themed challenges involve paradoxes about probability, time, and change; metapuzzles; and self-referentiality. Nineteen chapters advance in difficulty from relatively simple to highly complex. 1982 edition. 240pp. 5 3/8 x 8 1/2. 0-486-47027-X

SATAN, CANTOR AND INFINITY: Mind-Boggling Puzzles, Raymond M. Smullyan. A renowned mathematician tells stories of knights and knaves in an entertaining look at the logical precepts behind infinity, probability, time, and change. Requires a strong background in mathematics. Complete solutions. 288pp. 5 3/8 x 8 1/2. 0-486-47036-9

THE RED BOOK OF MATHEMATICAL PROBLEMS, Kenneth S. Williams and Kenneth Hardy. Handy compilation of 100 practice problems, hints and solutions indispensable for students preparing for the William Lowell Putnam and other mathematical competitions. Preface to the First Edition. Sources. 1988 edition. 192pp. 5 3/8 x 8 1/2. 0-486-69415-1

KING ARTHUR IN SEARCH OF HIS DOG AND OTHER CURIOUS PUZZLES, Raymond M. Smullyan. This fanciful, original collection for readers of all ages features arithmetic puzzles, logic problems related to crime detection, and logic and arithmetic puzzles involving King Arthur and his Dogs of the Round Table. 160pp. 5 3/8 x 8 1/2. 0-486-47435-6

UNDECIDABLE THEORIES: Studies in Logic and the Foundation of Mathematics, Alfred Tarski in collaboration with Andrzej Mostowski and Raphael M. Robinson. This well-known book by the famed logician consists of three treatises: "A General Method in Proofs of Undecidability," "Undecidability and Essential Undecidability in Mathematics," and "Undecidability of the Elementary Theory of Groups." 1953 edition. 112pp. 5 3/8 x 8 1/2. 0-486-47703-7

LOGIC FOR MATHEMATICIANS, J. Barkley Rosser. Examination of essential topics and theorems assumes no background in logic. "Undoubtedly a major addition to the literature of mathematical logic." – *Bulletin of the American Mathematical Society.* 1978 edition. 592pp. 6 1/8 x 9 1/4. 0-486-46898-4

INTRODUCTION TO PROOF IN ABSTRACT MATHEMATICS, Andrew Wohlgemuth. This undergraduate text teaches students what constitutes an acceptable proof, and it develops their ability to do proofs of routine problems as well as those requiring creative insights. 1990 edition. 384pp. 6 1/2 x 9 1/4. 0-486-47854-8

FIRST COURSE IN MATHEMATICAL LOGIC, Patrick Suppes and Shirley Hill. Rigorous introduction is simple enough in presentation and context for wide range of students. Symbolizing sentences; logical inference; truth and validity; truth tables; terms, predicates, universal quantifiers; universal specification and laws of identity; more. 288pp. 5 3/8 x 8 1/2. 0-486-42259-3

Mathematics-Algebra and Calculus

VECTOR CALCULUS, Peter Baxandall and Hans Liebeck. This introductory text offers a rigorous, comprehensive treatment. Classical theorems of vector calculus are amply illustrated with figures, worked examples, physical applications, and exercises with hints and answers. 1986 edition. 560pp. 5 3/8 x 8 1/2. 0-486-46620-5

ADVANCED CALCULUS: An Introduction to Classical Analysis, Louis Brand. A course in analysis that focuses on the functions of a real variable, this text introduces the basic concepts in their simplest setting and illustrates its teachings with numerous examples, theorems, and proofs. 1955 edition. 592pp. 5 3/8 x 8 1/2. 0-486-44548-8

ADVANCED CALCULUS, Avner Friedman. Intended for students who have already completed a one-year course in elementary calculus, this two-part treatment advances from functions of one variable to those of several variables. Solutions. 1971 edition. 432pp. 5 3/8 x 8 1/2. 0-486-45795-8

METHODS OF MATHEMATICS APPLIED TO CALCULUS, PROBABILITY, AND STATISTICS, Richard W. Hamming. This 4-part treatment begins with algebra and analytic geometry and proceeds to an exploration of the calculus of algebraic functions and transcendental functions and applications. 1985 edition. Includes 310 figures and 18 tables. 880pp. 6 1/2 x 9 1/4. 0-486-43945-3

BASIC ALGEBRA I: Second Edition, Nathan Jacobson. A classic text and standard reference for a generation, this volume covers all undergraduate algebra topics, including groups, rings, modules, Galois theory, polynomials, linear algebra, and associative algebra. 1985 edition. 528pp. 6 1/8 x 9 1/4. 0-486-47189-6

BASIC ALGEBRA II: Second Edition, Nathan Jacobson. This classic text and standard reference comprises all subjects of a first-year graduate-level course, including in-depth coverage of groups and polynomials and extensive use of categories and functors. 1989 edition. 704pp. 6 1/8 x 9 1/4. 0-486-47187-X

CALCULUS: An Intuitive and Physical Approach (Second Edition), Morris Kline. Application-oriented introduction relates the subject as closely as possible to science with explorations of the derivative; differentiation and integration of the powers of x; theorems on differentiation, antidifferentiation; the chain rule; trigonometric functions; more. Examples. 1967 edition. 960pp. 6 1/2 x 9 1/4. 0-486-40453-6

ABSTRACT ALGEBRA AND SOLUTION BY RADICALS, John E. Maxfield and Margaret W. Maxfield. Accessible advanced undergraduate-level text starts with groups, rings, fields, and polynomials and advances to Galois theory, radicals and roots of unity, and solution by radicals. Numerous examples, illustrations, exercises, appendixes. 1971 edition. 224pp. 6 1/8 x 9 1/4. 0-486-47723-1

AN INTRODUCTION TO THE THEORY OF LINEAR SPACES, Georgi E. Shilov. Translated by Richard A. Silverman. Introductory treatment offers a clear exposition of algebra, geometry, and analysis as parts of an integrated whole rather than separate subjects. Numerous examples illustrate many different fields, and problems include hints or answers. 1961 edition. 320pp. 5 3/8 x 8 1/2. 0-486-63070-6

LINEAR ALGEBRA, Georgi E. Shilov. Covers determinants, linear spaces, systems of linear equations, linear functions of a vector argument, coordinate transformations, the canonical form of the matrix of a linear operator, bilinear and quadratic forms, and more. 387pp. 5 3/8 x 8 1/2. 0-486-63518-X

Mathematics–Probability and Statistics

BASIC PROBABILITY THEORY, Robert B. Ash. This text emphasizes the probabilistic way of thinking, rather than measure-theoretic concepts. Geared toward advanced undergraduates and graduate students, it features solutions to some of the problems. 1970 edition. 352pp. 5 3/8 x 8 1/2. 0-486-46628-0

PRINCIPLES OF STATISTICS, M. G. Bulmer. Concise description of classical statistics, from basic dice probabilities to modern regression analysis. Equal stress on theory and applications. Moderate difficulty; only basic calculus required. Includes problems with answers. 252pp. 5 5/8 x 8 1/4. 0-486-63760-3

OUTLINE OF BASIC STATISTICS: Dictionary and Formulas, John E. Freund and Frank J. Williams. Handy guide includes a 70-page outline of essential statistical formulas covering grouped and ungrouped data, finite populations, probability, and more, plus over 1,000 clear, concise definitions of statistical terms. 1966 edition. 208pp. 5 3/8 x 8 1/2. 0-486-47769-X

GOOD THINKING: The Foundations of Probability and Its Applications, Irving J. Good. This in-depth treatment of probability theory by a famous British statistician explores Keynesian principles and surveys such topics as Bayesian rationality, corroboration, hypothesis testing, and mathematical tools for induction and simplicity. 1983 edition. 352pp. 5 3/8 x 8 1/2. 0-486-47438-0

INTRODUCTION TO PROBABILITY THEORY WITH CONTEMPORARY APPLICATIONS, Lester L. Helms. Extensive discussions and clear examples, written in plain language, expose students to the rules and methods of probability. Exercises foster problem-solving skills, and all problems feature step-by-step solutions. 1997 edition. 368pp. 6 1/2 x 9 1/4. 0-486-47418-6

CHANCE, LUCK, AND STATISTICS, Horace C. Levinson. In simple, non-technical language, this volume explores the fundamentals governing chance and applies them to sports, government, and business. "Clear and lively ... remarkably accurate." – *Scientific Monthly.* 384pp. 5 3/8 x 8 1/2. 0-486-41997-5

FIFTY CHALLENGING PROBLEMS IN PROBABILITY WITH SOLUTIONS, Frederick Mosteller. Remarkable puzzlers, graded in difficulty, illustrate elementary and advanced aspects of probability. These problems were selected for originality, general interest, or because they demonstrate valuable techniques. Also includes detailed solutions. 88pp. 5 3/8 x 8 1/2. 0-486-65355-2

EXPERIMENTAL STATISTICS, Mary Gibbons Natrella. A handbook for those seeking engineering information and quantitative data for designing, developing, constructing, and testing equipment. Covers the planning of experiments, the analyzing of extreme-value data; and more. 1966 edition. Index. Includes 52 figures and 76 tables. 560pp. 8 3/8 x 11. 0-486-43937-2

STOCHASTIC MODELING: Analysis and Simulation, Barry L. Nelson. Coherent introduction to techniques also offers a guide to the mathematical, numerical, and simulation tools of systems analysis. Includes formulation of models, analysis, and interpretation of results. 1995 edition. 336pp. 6 1/8 x 9 1/4. 0-486-47770-3

INTRODUCTION TO BIOSTATISTICS: Second Edition, Robert R. Sokal and F. James Rohlf. Suitable for undergraduates with a minimal background in mathematics, this introduction ranges from descriptive statistics to fundamental distributions and the testing of hypotheses. Includes numerous worked-out problems and examples. 1987 edition. 384pp. 6 1/8 x 9 1/4. 0-486-46961-1

Mathematics-Geometry and Topology

PROBLEMS AND SOLUTIONS IN EUCLIDEAN GEOMETRY, M. N. Aref and William Wernick. Based on classical principles, this book is intended for a second course in Euclidean geometry and can be used as a refresher. More than 200 problems include hints and solutions. 1968 edition. 272pp. 5 3/8 x 8 1/2. 0-486-47720-7

TOPOLOGY OF 3-MANIFOLDS AND RELATED TOPICS, Edited by M. K. Fort, Jr. With a New Introduction by Daniel Silver. Summaries and full reports from a 1961 conference discuss decompositions and subsets of 3-space; n-manifolds; knot theory; the Poincaré conjecture; and periodic maps and isotopies. Familiarity with algebraic topology required. 1962 edition. 272pp. 6 1/8 x 9 1/4. 0-486-47753-3

POINT SET TOPOLOGY, Steven A. Gaal. Suitable for a complete course in topology, this text also functions as a self-contained treatment for independent study. Additional enrichment materials make it equally valuable as a reference. 1964 edition. 336pp. 5 3/8 x 8 1/2. 0-486-47222-1

INVITATION TO GEOMETRY, Z. A. Melzak. Intended for students of many different backgrounds with only a modest knowledge of mathematics, this text features self-contained chapters that can be adapted to several types of geometry courses. 1983 edition. 240pp. 5 3/8 x 8 1/2. 0-486-46626-4

TOPOLOGY AND GEOMETRY FOR PHYSICISTS, Charles Nash and Siddhartha Sen. Written by physicists for physics students, this text assumes no detailed background in topology or geometry. Topics include differential forms, homotopy, homology, cohomology, fiber bundles, connection and covariant derivatives, and Morse theory. 1983 edition. 320pp. 5 3/8 x 8 1/2. 0-486-47852-1

BEYOND GEOMETRY: Classic Papers from Riemann to Einstein, Edited with an Introduction and Notes by Peter Pesic. This is the only English-language collection of these 8 accessible essays. They trace seminal ideas about the foundations of geometry that led to Einstein's general theory of relativity. 224pp. 6 1/8 x 9 1/4. 0-486-45350-2

GEOMETRY FROM EUCLID TO KNOTS, Saul Stahl. This text provides a historical perspective on plane geometry and covers non-neutral Euclidean geometry, circles and regular polygons, projective geometry, symmetries, inversions, informal topology, and more. Includes 1,000 practice problems. Solutions available. 2003 edition. 480pp. 6 1/8 x 9 1/4. 0-486-47459-3

TOPOLOGICAL VECTOR SPACES, DISTRIBUTIONS AND KERNELS, François Trèves. Extending beyond the boundaries of Hilbert and Banach space theory, this text focuses on key aspects of functional analysis, particularly in regard to solving partial differential equations. 1967 edition. 592pp. 5 3/8 x 8 1/2. 0-486-45352-9

INTRODUCTION TO PROJECTIVE GEOMETRY, C. R. Wylie, Jr. This introductory volume offers strong reinforcement for its teachings, with detailed examples and numerous theorems, proofs, and exercises, plus complete answers to all odd-numbered end-of-chapter problems. 1970 edition. 576pp. 6 1/8 x 9 1/4. 0-486-46895-X

FOUNDATIONS OF GEOMETRY, C. R. Wylie, Jr. Geared toward students preparing to teach high school mathematics, this text explores the principles of Euclidean and non-Euclidean geometry and covers both generalities and specifics of the axiomatic method. 1964 edition. 352pp. 6 x 9. 0-486-47214-0

Mathematics-History

THE WORKS OF ARCHIMEDES, Archimedes. Translated by Sir Thomas Heath. Complete works of ancient geometer feature such topics as the famous problems of the ratio of the areas of a cylinder and an inscribed sphere; the properties of conoids, spheroids, and spirals; more. 326pp. 5 3/8 x 8 1/2. 0-486-42084-1

THE HISTORICAL ROOTS OF ELEMENTARY MATHEMATICS, Lucas N. H. Bunt, Phillip S. Jones, and Jack D. Bedient. Exciting, hands-on approach to understanding fundamental underpinnings of modern arithmetic, algebra, geometry and number systems examines their origins in early Egyptian, Babylonian, and Greek sources. 336pp. 5 3/8 x 8 1/2. 0-486-25563-8

THE THIRTEEN BOOKS OF EUCLID'S ELEMENTS, Euclid. Contains complete English text of all 13 books of the Elements plus critical apparatus analyzing each definition, postulate, and proposition in great detail. Covers textual and linguistic matters; mathematical analyses of Euclid's ideas; classical, medieval, Renaissance and modern commentators; refutations, supports, extrapolations, reinterpretations and historical notes. 995 figures. Total of 1,425pp. All books 5 3/8 x 8 1/2.
Vol. I: 443pp. 0-486-60088-2
Vol. II: 464pp. 0-486-60089-0
Vol. III: 546pp. 0-486-60090-4

A HISTORY OF GREEK MATHEMATICS, Sir Thomas Heath. This authoritative two-volume set that covers the essentials of mathematics and features every landmark innovation and every important figure, including Euclid, Apollonius, and others. 5 3/8 x 8 1/2. Vol. I: 461pp. 0-486-24073-8
Vol. II: 597pp. 0-486-24074-6

A MANUAL OF GREEK MATHEMATICS, Sir Thomas L. Heath. This concise but thorough history encompasses the enduring contributions of the ancient Greek mathematicians whose works form the basis of most modern mathematics. Discusses Pythagorean arithmetic, Plato, Euclid, more. 1931 edition. 576pp. 5 3/8 x 8 1/2.
0-486-43231-9

CHINESE MATHEMATICS IN THE THIRTEENTH CENTURY, Ulrich Libbrecht. An exploration of the 13th-century mathematician Ch'in, this fascinating book combines what is known of the mathematician's life with a history of his only extant work, the Shu-shu chiu-chang. 1973 edition. 592pp. 5 3/8 x 8 1/2.
0-486-44619-0

PHILOSOPHY OF MATHEMATICS AND DEDUCTIVE STRUCTURE IN EUCLID'S ELEMENTS, Ian Mueller. This text provides an understanding of the classical Greek conception of mathematics as expressed in Euclid's Elements. It focuses on philosophical, foundational, and logical questions and features helpful appendixes. 400pp. 6 1/2 x 9 1/4. 0-486-45300-6

BEYOND GEOMETRY: Classic Papers from Riemann to Einstein, Edited with an Introduction and Notes by Peter Pesic. This is the only English-language collection of these 8 accessible essays. They trace seminal ideas about the foundations of geometry that led to Einstein's general theory of relativity. 224pp. 6 1/8 x 9 1/4. 0-486-45350-2

HISTORY OF MATHEMATICS, David E. Smith. Two-volume history – from Egyptian papyri and medieval maps to modern graphs and diagrams. Non-technical chronological survey with thousands of biographical notes, critical evaluations, and contemporary opinions on over 1,100 mathematicians. 5 3/8 x 8 1/2.
Vol. I: 618pp. 0-486-20429-4
Vol. II: 736pp. 0-486-20430-8

Physics

THEORETICAL NUCLEAR PHYSICS, John M. Blatt and Victor F. Weisskopf. An uncommonly clear and cogent investigation and correlation of key aspects of theoretical nuclear physics by leading experts: the nucleus, nuclear forces, nuclear spectroscopy, two-, three- and four-body problems, nuclear reactions, beta-decay and nuclear shell structure. 896pp. 5 3/8 x 8 1/2. 0-486-66827-4

QUANTUM THEORY, David Bohm. This advanced undergraduate-level text presents the quantum theory in terms of qualitative and imaginative concepts, followed by specific applications worked out in mathematical detail. 655pp. 5 3/8 x 8 1/2. 0-486-65969-0

ATOMIC PHYSICS AND HUMAN KNOWLEDGE, Niels Bohr. Articles and speeches by the Nobel Prize–winning physicist, dating from 1934 to 1958, offer philosophical explorations of the relevance of atomic physics to many areas of human endeavor. 1961 edition. 112pp. 5 3/8 x 8 1/2. 0-486-47928-5

COSMOLOGY, Hermann Bondi. A co-developer of the steady-state theory explores his conception of the expanding universe. This historic book was among the first to present cosmology as a separate branch of physics. 1961 edition. 192pp. 5 3/8 x 8 1/2. 0-486-47483-6

LECTURES ON QUANTUM MECHANICS, Paul A. M. Dirac. Four concise, brilliant lectures on mathematical methods in quantum mechanics from Nobel Prize-winning quantum pioneer build on idea of visualizing quantum theory through the use of classical mechanics. 96pp. 5 3/8 x 8 1/2. 0-486-41713-1

THE PRINCIPLE OF RELATIVITY, Albert Einstein and Frances A. Davis. Eleven papers that forged the general and special theories of relativity include seven papers by Einstein, two by Lorentz, and one each by Minkowski and Weyl. 1923 edition. 240pp. 5 3/8 x 8 1/2. 0-486-60081-5

PHYSICS OF WAVES, William C. Elmore and Mark A. Heald. Ideal as a classroom text or for individual study, this unique one-volume overview of classical wave theory covers wave phenomena of acoustics, optics, electromagnetic radiations, and more. 477pp. 5 3/8 x 8 1/2. 0-486-64926-1

THERMODYNAMICS, Enrico Fermi. In this classic of modern science, the Nobel Laureate presents a clear treatment of systems, the First and Second Laws of Thermodynamics, entropy, thermodynamic potentials, and much more. Calculus required. 160pp. 5 3/8 x 8 1/2. 0-486-60361-X

QUANTUM THEORY OF MANY-PARTICLE SYSTEMS, Alexander L. Fetter and John Dirk Walecka. Self-contained treatment of nonrelativistic many-particle systems discusses both formalism and applications in terms of ground-state (zero-temperature) formalism, finite-temperature formalism, canonical transformations, and applications to physical systems. 1971 edition. 640pp. 5 3/8 x 8 1/2. 0-486-42827-3

QUANTUM MECHANICS AND PATH INTEGRALS: Emended Edition, Richard P. Feynman and Albert R. Hibbs. Emended by Daniel F. Styer. The Nobel Prize–winning physicist presents unique insights into his theory and its applications. Feynman starts with fundamentals and advances to the perturbation method, quantum electrodynamics, and statistical mechanics. 1965 edition, emended in 2005. 384pp. 6 1/8 x 9 1/4. 0-486-47722-3

Physics

INTRODUCTION TO MODERN OPTICS, Grant R. Fowles. A complete basic undergraduate course in modern optics for students in physics, technology, and engineering. The first half deals with classical physical optics; the second, quantum nature of light. Solutions. 336pp. 5 3/8 x 8 1/2. 0-486-65957-7

THE QUANTUM THEORY OF RADIATION: Third Edition, W. Heitler. The first comprehensive treatment of quantum physics in any language, this classic introduction to basic theory remains highly recommended and widely used, both as a text and as a reference. 1954 edition. 464pp. 5 3/8 x 8 1/2. 0-486-64558-4

QUANTUM FIELD THEORY, Claude Itzykson and Jean-Bernard Zuber. This comprehensive text begins with the standard quantization of electrodynamics and perturbative renormalization, advancing to functional methods, relativistic bound states, broken symmetries, nonabelian gauge fields, and asymptotic behavior. 1980 edition. 752pp. 6 1/2 x 9 1/4. 0-486-44568-2

FOUNDATIONS OF POTENTIAL THERY, Oliver D. Kellogg. Introduction to fundamentals of potential functions covers the force of gravity, fields of force, potentials, harmonic functions, electric images and Green's function, sequences of harmonic functions, fundamental existence theorems, and much more. 400pp. 5 3/8 x 8 1/2. 0-486-60144-7

FUNDAMENTALS OF MATHEMATICAL PHYSICS, Edgar A. Kraut. Indispensable for students of modern physics, this text provides the necessary background in mathematics to study the concepts of electromagnetic theory and quantum mechanics. 1967 edition. 480pp. 6 1/2 x 9 1/4. 0-486-45809-1

GEOMETRY AND LIGHT: The Science of Invisibility, Ulf Leonhardt and Thomas Philbin. Suitable for advanced undergraduate and graduate students of engineering, physics, and mathematics and scientific researchers of all types, this is the first authoritative text on invisibility and the science behind it. More than 100 full-color illustrations, plus exercises with solutions. 2010 edition. 288pp. 7 x 9 1/4. 0-486-47693-6

QUANTUM MECHANICS: New Approaches to Selected Topics, Harry J. Lipkin. Acclaimed as "excellent" (*Nature*) and "very original and refreshing" (*Physics Today*), these studies examine the Mössbauer effect, many-body quantum mechanics, scattering theory, Feynman diagrams, and relativistic quantum mechanics. 1973 edition. 480pp. 5 3/8 x 8 1/2. 0-486-45893-8

THEORY OF HEAT, James Clerk Maxwell. This classic sets forth the fundamentals of thermodynamics and kinetic theory simply enough to be understood by beginners, yet with enough subtlety to appeal to more advanced readers, too. 352pp. 5 3/8 x 8 1/2. 0-486-41735-2

QUANTUM MECHANICS, Albert Messiah. Subjects include formalism and its interpretation, analysis of simple systems, symmetries and invariance, methods of approximation, elements of relativistic quantum mechanics, much more. "Strongly recommended." – *American Journal of Physics.* 1152pp. 5 3/8 x 8 1/2. 0-486-40924-4

RELATIVISTIC QUANTUM FIELDS, Charles Nash. This graduate-level text contains techniques for performing calculations in quantum field theory. It focuses chiefly on the dimensional method and the renormalization group methods. Additional topics include functional integration and differentiation. 1978 edition. 240pp. 5 3/8 x 8 1/2. 0-486-47752-5

Physics

MATHEMATICAL TOOLS FOR PHYSICS, James Nearing. Encouraging students' development of intuition, this original work begins with a review of basic mathematics and advances to infinite series, complex algebra, differential equations, Fourier series, and more. 2010 edition. 496pp. 6 1/8 x 9 1/4. 0-486-48212-X

TREATISE ON THERMODYNAMICS, Max Planck. Great classic, still one of the best introductions to thermodynamics. Fundamentals, first and second principles of thermodynamics, applications to special states of equilibrium, more. Numerous worked examples. 1917 edition. 297pp. 5 3/8 x 8. 0-486-66371-X

AN INTRODUCTION TO RELATIVISTIC QUANTUM FIELD THEORY, Silvan S. Schweber. Complete, systematic, and self-contained, this text introduces modern quantum field theory. "Combines thorough knowledge with a high degree of didactic ability and a delightful style." – *Mathematical Reviews.* 1961 edition. 928pp. 5 3/8 x 8 1/2. 0-486-44228-4

THE ELECTROMAGNETIC FIELD, Albert Shadowitz. Comprehensive undergraduate text covers basics of electric and magnetic fields, building up to electromagnetic theory. Related topics include relativity theory. Over 900 problems, some with solutions. 1975 edition. 768pp. 5 5/8 x 8 1/4. 0-486-65660-8

THE PRINCIPLES OF STATISTICAL MECHANICS, Richard C. Tolman. Definitive treatise offers a concise exposition of classical statistical mechanics and a thorough elucidation of quantum statistical mechanics, plus applications of statistical mechanics to thermodynamic behavior. 1930 edition. 704pp. 5 5/8 x 8 1/4. 0-486-63896-0

INTRODUCTION TO THE PHYSICS OF FLUIDS AND SOLIDS, James S. Trefil. This interesting, informative survey by a well-known science author ranges from classical physics and geophysical topics, from the rings of Saturn and the rotation of the galaxy to underground nuclear tests. 1975 edition. 320pp. 5 3/8 x 8 1/2. 0-486-47437-2

STATISTICAL PHYSICS, Gregory H. Wannier. Classic text combines thermodynamics, statistical mechanics, and kinetic theory in one unified presentation. Topics include equilibrium statistics of special systems, kinetic theory, transport coefficients, and fluctuations. Problems with solutions. 1966 edition. 532pp. 5 3/8 x 8 1/2. 0-486-65401-X

SPACE, TIME, MATTER, Hermann Weyl. Excellent introduction probes deeply into Euclidean space, Riemann's space, Einstein's general relativity, gravitational waves and energy, and laws of conservation. "A classic of physics." – *British Journal for Philosophy and Science.* 330pp. 5 3/8 x 8 1/2. 0-486-60267-2

RANDOM VIBRATIONS: Theory and Practice, Paul H. Wirsching, Thomas L. Paez and Keith Ortiz. Comprehensive text and reference covers topics in probability, statistics, and random processes, plus methods for analyzing and controlling random vibrations. Suitable for graduate students and mechanical, structural, and aerospace engineers. 1995 edition. 464pp. 5 3/8 x 8 1/2. 0-486-45015-5

PHYSICS OF SHOCK WAVES AND HIGH-TEMPERATURE HYDRO DYNAMIC PHENOMENA, Ya B. Zel'dovich and Yu P. Raizer. Physical, chemical processes in gases at high temperatures are focus of outstanding text, which combines material from gas dynamics, shock-wave theory, thermodynamics and statistical physics, other fields. 284 illustrations. 1966–1967 edition. 944pp. 6 1/8 x 9 1/4. 0-486-42002-7